Brunelleschi
*Studies of His Technology
and Inventions*

·D·S·
PHILIPPVS ARCHITECTVS

Brunelleschi

Studies of His Technology and Inventions

FRANK D. PRAGER

and

GUSTINA SCAGLIA

Dover Publications, Inc.
Mineola, New York

Bibliographical Note

This Dover edition, first published in 2004, is an unabridged republication of the work originally published by The MIT Press, Cambridge, Massachusetts, in 1970. This edition has been published by special arrangement with The MIT Press.

Library of Congress Cataloging-in-Publication Data

Prager, Frank D.
    Brunelleschi : studies of his technology and inventions / Frank D. Prager and Gustina Scaglia.
        p. cm.
    Includes bibliographical references and index.
    ISBN-13: 978-0-486-43464-3 (pbk.)
    ISBN-10: 0-486-43464-8 (pbk.)
    1. Brunelleschi, Filippo, 1377–1446—Criticism and interpretation. 2. Building—Italy—Florence—History—15th century. I. Scaglia, Gustina. II. Title.

NA1123.B8P7 2004
720'.92—dc22

2004049421

Manufactured in the United States by RR Donnelley
43464804   2016
www.doverpublications.com

W hen we undertook these studies, we knew that the architectural works of Filippo Brunelleschi belong to the most famous and the most controversial creations of art. We noted that his methods are equally renowned and equally questioned in the history of technology. Admirers of Filippo are convinced that he, single-handedly, transformed both building art and building technique by fundamental inventions. Others rather see him as simply reacting to earlier trends or innovations.

Vasari, a historian unusually well acquainted with Filippo, described him as one of the men "who, though puny in person and insignificant of figure, are yet endowed with so much greatness of soul and such force of character that they, unless they can occupy themselves with difficult, nay, almost impossible undertakings and carry them to perfection, they can find no peace in their lives." Such men are always controversial, and we were not surprised to find that the Dome officials for whom Filippo worked repeatedly criticized his human reactions. Even his friends, Taccola and Manetti, show him as irascible, sarcastic, and self-centered. It may still be true that Filippo had unique powers of visualization and visual expression and that he neither wanted nor needed assistance. In our studies we were interested in these problems and were prepared to find support for any of the views in print.

However, if Filippo left an imprint nearly as deep as is often assumed, we hoped to learn by what incidents or powers he achieved it.

The authentic structures are most fascinating, and the records are both plentiful and interesting, although they are complex and often inconclusive. Even less conclusive are the evaluations of scholars. Successive generations see Filippo through different eyes. Historians have found in him an heir to Gothic achievement, a victor over Gothic barbarism, or the exponent of a permanent Classic presence, either a pure artist or a mathematical genius, and even—if possible—something of all these things.

We tried to determine authentic facts about Filippo and his works and tried this on a fairly comprehensive basis. He was an engineer, an artist, a businessman, and sometimes a man active in public affairs. For these reasons we took an interest in observing, if possible, his position in technology, art, economy, and history. We would be equally interested in views that he may have held in science, philosophy, ethics, and literature, but we found almost nothing reported that he said or wrote about such matters, and we thought it would be futile to speculate about his possible attitudes in matters of such complexity. On the other hand, we found his technical work, or in some cases at least his technical possibilities, capable of description in relatively definite terms in view of the simple and rigid laws of mechanics.

The studies presented here are a revised and unified edition of articles originally published by Prager as contributions to the history of inventions and technology, and by Scaglia as work in the history of artists and art. The present revision is formulated by Prager, although it is based on review by both of us. We received help from a great number of our friends, and we had the assistance of capable librarians in Florence and elsewhere. We are specially indebted even now to the late George Sarton. Valuable information was given by Ulrich Middeldorf and Alessandro Parronchi. The drawings outlined for this book were capably and patiently prepared by George Rich of Philadelphia. Last but not least, our thanks go to Henry Millon, architect and historian at M.I.T., for suggesting this new edition of our studies and for aiding us by his kind and careful review of an earlier draft.

FRANK D. PRAGER

GUSTINA SCAGLIA

October 1969

WORKS FREQUENTLY CITED

Brunelleschiana
C. von Fabriczy, *Brunelleschiana*, Berlin, 1907.

Costruzione
C. Guasti, *Santa Maria del Fiore. La costruzione*, Florence, 1887.

Cupola
C. Guasti, *La cupola di Santa Maria del Fiore*, Florence, 1857.

Fabriczy
C. von Fabriczy, *Filippo Brunelleschi, sein Leben und seine Werke*, Stuttgart, 1892.

Forma e colore
P. Sanpaolesi, *La cupola del Brunelleschi* (Forma e colore. I grandi cicli dell'arte), Florence, 1965.

Il progetto
P. Sanpaolesi, *La cupola di Santa Maria del Fiore. Il progetto. La costruzione*, Rome, 1941.

Manetti
A. Manetti, *Vita di Filippo Brunelleschi*, ed. E. Toesca, Florence, 1927.

Nardini
A. Nardini-Despotti-Mospignotti, *Filippo Brunelleschi e la cupola*, Florence, 1885.

Nelli
G. B. Nelli, *Piante ed alzati interiori . . . di S. Maria del Fiore . . .* , ed. G. B. C. Nelli, Florence, 1755.

Rilievi
Opera di Santa Maria del Fiore, *Rilievi e studi sulla cupola del Brunelleschi*, Florence, 1939.

Sanpaolesi
P. Sanpaolesi, *Brunelleschi*, Milan, 1962.

Trattati
Francesco di Giorgio, *Trattati*, ed. C. Maltese, Milan, 1967.

Vasari
G. Vasari, *Le vite de' più eccellenti architettori, pittori e scultori italiani*, 1568, translated by G. du C. de Vere, London, 1912–1915.

Zibaldone
Buonaccorso Ghiberti, *Zibaldone*, unpublished manuscript, BR 228, Florence, Biblioteca Nazionale.

The Florentine *braccio* was 23 inches long.[2] When the documents specify architectural measures, we generally translate 1 *braccio* as 2 feet, but 10 *braccia* as 19 feet, thus rounding the measure up or down to the nearest integral number of feet, except in the rare cases where the text specifies fractions.[3]

The *libra* equaled slightly less than ¾ pound avoirdupois,[4] which we call ¾ English pound. A ton, *migliaia*, was 1,000 *libre* and a *libra* had 12 *unciae*.

The *lira* had 20 *soldi* of 12 *denarii* each.[5] The *florin* was worth 3.13 to 3.15 *lire* in Brunelleschi's time.[6]

Typical wages and salaries were as follows:[7]

Common laborers	10	*soldi* per day
Master carpenters	1	*lira* per day
Master bricklayers	1	*lira* per day
Mercantile agents	100	*lire* per year
Notaries, up to	400	*lire* per year

The Calendar used the Annunciation system. For example the Florentine year 1420 ran from "25 March 1420" (our 25 March 1420) to "24 March 1420" (our 24 March *1421*).[8] In our translations we generally use the modern form of stating the date. In some cases, where this could lead to confusion or doubt, we give both dates, for example, "24 March 1420–1421."

[7] G. P. Pagnini, *Della Decima*, Florence, 1765, I, pp. 116, 263 ff., Pl. I; R. Davidsohn, *Geschichte von Florenz*, Berlin, 1897–1927, II, ii, pp. 402 ff.; IV, ii, pp. 208 ff.; also see J. E. Staley, *The Guilds of Florence*, London, 1906, pp. 92 f., 153, 320.

[8] F. K. Ginzel, *Handbuch der mathematischen und technischen Chronologie. Das Zeitrechnungswesen der Völker*, III, Leipzig, 1914, pp. 161 ff.

[1] Only approximate equivalents of the old standards are used here. These standards, derived from such approximate reference data as the human arm and foot, had no precise relation to one another. Therefore, no precise relation can be made with modern values divisible into small fractions. For comparative lists of measures used in different parts of Italy, see the Appendix of W. B. Parsons, *Engineers and Engineering in the Renaissance*, Baltimore, 1939.

[2] See also the introduction to *Santa Maria del Fiore. La Costruzione*, ed. C. Guasti, Florence, 1887, p. LIX (hereafter: *Costruzione*).

[3] W. B. Scaife, *Florentine Life during the Renaissance*, Baltimore, 1893, p. 175, gives 22.97 inches. This is mere pseudoprecision.

[4] A. Schaube, *Handelsgeschichte der romanischen Völker*, 1906, p. 814.

[5] The proportion 20 × 12 survives in England. The words *lira, sou*, and *dinar* survive elsewhere.

[6] Its value was frequently reestablished by legislation. It weighed 3.54 grams and contained 985/1000 fine gold. The obverse showed the Florentine lily, the verso St. John.

Most of the facts collected in the following note were discovered by Gustina Scaglia, together with persuasive proofs of attributions and dates. The facts—stated here without the proofs, which would lead us too far at this point—suggest certain conclusions relating to the work of Brunelleschi, Ghiberti, Francesco di Giorgio, Leonardo, and others. While we may not agree on each detail of the conclusions, we agree on the following general survey of our source materials recorded on paper or parchment.

Brunelleschi rarely resorted to the use of paper and the like, but his rival Lorenzo Ghiberti and his friend Mariano Taccola left us a record of his technology. Ghiberti's collection or *Zibaldone* may have been redrawn and rewritten by his grandson Buonaccorso; fortunately it survives (Florence, Biblioteca Nazionale, BR 228). This set does not mention Brunelleschi's name but is close enough to the more explicit Taccola and to the still more specific documents of the Opera del Duomo to pass as a semiauthentic record of Filippo's ideas, at least with reference to mechanics.

Taccola wrote a treatise *De ingeneis*, finished in 1433, about the time that Brunelleschi finished the Cupola. (Books I and II are in Munich, Lat. 197; III and IV are in Florence, Biblioteca Nazionale, Palat. 766.) In a Sequel to this work (Munich) Taccola writes specifically and pungently about Brunelleschi and quotes a long speech of the master. Taccola later rewrote his treatise, in his *De machinis* (Munich, Lat. 28800), and that version as well as the first was copied repeatedly; there are at least three copies of the later work (New York, Spencer 136; Paris, Bibliothèque Nationale, Lat. 7239; Venice, Marciana, Lat. 2941, the latter being mixed with other copywork). There is no evidence that anyone of significance for the history of technology read the later work or any of its copies, which were made for men of war and of state. There is, however, much evidence that Taccola's earlier work was read and studied by men of great technical importance. Selections from his characteristic graphic and written expressions—including many of those apparently reflecting Brunelleschian ideas—recur unmistakably in works of Francesco di Giorgio (who also added a few notes to Lat. 197), of Antonio da San Gallo, Bartolommeo Neroni, Oreste Biringuccio, and others, as will be described in the following paragraphs.

Almost invariably these selections from Brunelleschi-Taccola are mixed with copywork from a Machinery Complex of unknown parentage, which also appears to be strongly influenced by Brunelleschi. Its archetype has not been found thus far. It deals with wormgear drives, mainly for hoists, cars, and mills, and with other mill drives.

One of the first, and surely one of the principal works containing this mixture of elements, is the *Opusculum de architectura*, an autograph of Francesco di Giorgio datable about 1472 (London, British Museum, Harley 3281). This London manuscript is substantially limited to machinery, except for a few fortress plans, but is completed by another and more truly "architectural" autograph of Francesco di Giorgio, possibly begun when he finished his mechanical copywork. Fragments of the architectural record are preserved in Florence (Uffizi, *Taccuino* of Francesco di Giorgio). In addition the Vaticana in Rome (Urb. Lat. 1757) has a minute version of the London manuscript. It may be Francesco's personal record of his mechanical studies and developments.

A vastly amplified and clarified version of these beginnings is the substance of Francesco's *Trattati* (Florence, Laurentiana, ex-Ashburnham 361, and Turin, Biblioteca Reale, Saluzziano 148, both published by C. Maltese, Milan, 1967). Here the ex-pupil of Taccola presents among other things an entirely new text, original as well as careful in formulation, although fragments of the source materials can be detected. These Treatises are written by scribes, and we think the illustrations are drawn by hired draftsmen, but there are a few corrections and additions in Francesco's own hand. Brunelleschian forms of architecture may also be seen in the Treatises, side by side with rediscovered Roman architecture and Vitruvian theory. The work of Francesco also shows a newly developing architecture of fortresses; its history is often described but continues to pose questions, as the transformations of historic forms, shown in Francesco's

record, are manifold and complex.

Contemporaries of Francesco produced somewhat similar copywork, as to machines, but did not show nearly as much originality. This may be noted from a codex in Siena (Biblioteca Comunale, S.IV.5); one in Milan (Ambrosiana, N.191); and several in Florence, Biblioteca Nazionale (Magliabecchiano II.III.314 and XVIII.V.2 and Palat. 767, the latter also copied in Venice, Marciana, Ital. Z.86). All these have, mainly, selections from Taccola and the Machinery Complex.

One man of the late Quattrocento seemed to follow Brunelleschi without use of Taccola or Francesco di Giorgio as intermediaries. This was Giuliano da San Gallo. (His notebook, S.IV.8 in Siena, was published by R. Falb, *Il taccuino senese di Giuliano San Gallo*, Siena, 1902.) Again different is the position of his nephew and pupil Antonio the Younger, who again made many scores of exact copies from rather medieval drawings of Taccola (preserved in the Uffizi in Florence). The San Gallos also continued the historic architectural research, initiated by Brunelleschi. More than Brunelleschi and Francesco di Giorgio, they concentrated on details and ornaments to be found in the ruins. Generally they may be seen as belonging to the generation of the "cricket cage," ridiculed by Michelangelo.

This also was the time of Leonardo. It is unnecessary here to discuss this great man's fame as an artist, but it must be noted that he appears rather secondary as a follower of Brunelleschian technology, although he contributed to putting the uppermost dot (the *palla*) on the Cupola. His note-books show drawings of the famous hoist and load positioners, in positions similar to those shown by Ghiberti and Giuliano San Gallo, and without original contribution, comparable to those of Francesco di Giorgio.

The books of Francesco were copied again in the Cinquecento. An important copy is preserved in Florence (Accademia, E.2.I.28). It is followed by others (Florence, B.N., Magl. II.I.141, and Siena, B.C., S.IV.4), which are sometimes believed to be from Francesco's lifetime but more probably depend on work done decades after his death. No doubt his copyists and followers continued to delve into his famous Treatises, and the Florentine copy (II.I.141) includes, in misplaced position, a long autographic translation of Vitruvius, by Francesco. More definitive reviews of this period may be expected from G. Scaglia and A. Parronchi.

Still other versions of the Brunelleschian ideas, usually derived via Taccola or Francesco or both, were made in the Cinquecento by such men as Bartolommeo Neroni (Siena, Biblioteca Comunale, S.IV.6), Pietro Cataneo (Florence, Uffizi, *Libro di Cataneo*), and Oreste Biringuccio (Siena, Biblioteca Comunale, S.IV.1). At this late time the Brunelleschian elements are less abundant but can still be found on some of the manuscript pages. Gradually they also found their way into printed books on technology and architecture, although we looked for them in vain in the books of the renewers of Vitruvian architecture: Alberti, Serlio, Palladio, and others. As is well known, the master influenced the style of the Renaissance, but his name was not mentioned and his specific ideas were not considered in most of the written output of that period. He was partly forgotten, but it seems he was too strong to be forgotten entirely.

F.D.P.

Brunelleschi
*Studies of His Technology
and Inventions*

BEGINNINGS OF
BRUNELLESCHIAN CONCEPTS[1]

1 In his own day, Filippo Brunelle-schi (Florence, 1377–1446) became known as the man who "renewed Roman masonry work." This renewal is now identified with the architectural Renaissance. The sober-minded Floren-tines appreciated its "economy" as well as its "harmonious proportions," and they said its economy came from the introduction of "vaulting without arma-ture." Such was the view of Filippo's chief employer, the cathedral-building office of the guild of woolen manufac-turers, and also the view of his first biographer, the humanist Manetti.[2] More or less the same view recurs in an early expression of Alberti, a man outstanding among those who devel-oped the Brunelleschian style:

> Who is so dull or jealous that he would not admire Filippo the archi-tect, in the face of this gigantic build-ing, rising above the vaults of heaven, wide enough to receive in its shade all the people of Tuscany, and built without the aid of any trusswork or mass of timber. . . .[3]

It is odd to see the work of the famous renewer equated with nothing more than a money-saving expedient. It seems more plausible to read in Vasari, the author of the classic Life of Brunelle-schi, that the master created a new possibility, not only a saving of cost. The builders in Florence were worried, Vasari says, "that no way could ever be found . . . to make a bridge strong enough to sustain the weight." Filippo overwhelmed them by showing that one could build "so great an edifice" with-out this "bridge," not merely with a cheaper bridge. Vasari agrees with Manetti that it was the master's wish "to restore to light the good manner of architecture" and that he achieved this, thereby establishing "no less a name for himself than Cimabue and Giotto had done."[4]

Very different are the views prevailing today, which we will show and discuss throughout these studies. When all the evidence is considered, old Vasari may still be right in many respects. Modern archivists have discovered documents

[1] Studies 1 to 3 are a revised edition of F. D. Prager, "Brunelleschi's Inventions and the 'Renewal of Roman Masonry Work,'" Osiris, IX, 1950, pp. 457 ff.
[2] A. Manetti, Vita di Filippo Brunelleschi, ca. 1485, Florence, Biblioteca Nazionale, II. II. 325, ed. E. Toesca, pp. 2, 19 f., 31 (hereafter: Manetti). An English transla-tion of excerpts is offered by E. G. Holt, Documentary History of Art, I, New York, 1957. New editions of Manetti, both English and Italian, are in preparation. Also see La cupola di Santa Maria del Fiore, ed. C. Guasti, Florence, 1857, Doc. 119 (hereafter: Cupola). Manetti and Cu-pola remain the chief sources of informa-tion about Brunelleschi recorded in writ-ing, but a new and improved edition of Cupola is badly needed.

[3] L. B. Alberti, Della pittura, 1434, ed. L. Mallé, Florence, 1950. There is a trace of irony in Alberti's remark about the Cu-pola's "shade" as many Florentines, then as now, found the inside too dark.
[4] G. Vasari, Le vite de' più eccellenti arch-itetti, etc., ed. 1568. (We quote the En-glish translation of G. du C. de Vere, London, 1912–1915, Vol. II, hereafter: Vasari). For the point under consideration see pp. 202–208, 211. Vasari disregards the "savings" except for secondary reference to them in a fictitious speech, p. 207.

that disagree in some points with the biographers' views. They show that hero worship had been practiced. It does not follow that the biographers' views are totally wrong or that, for example, the innovations claimed for Filippo were anticipated in the Trecento. We will cite and consider such authentic evidence as we can find. Of course we will also note the explanations suggested by biographers and other reviewers.

BACKGROUND

Even before Filippo's birth a controversy about the Florentine cathedral had come before his father, Ser Brunellesco di Lippo Lapi,[5] and even before the time of Ser Brunellesco there had been major architectural debates relating to Santa Reparata, the older metropolitan church.

This church, which also served as a meeting place for popular referenda, had been founded at the approximate location of the present *Duomo* in or about the seventh century.[6] Six

[5] *Santa Maria del Fiore. La costruzione*, ed. C. Guasti, Florence, 1887, Doc. 190 at p. 200 hereafter: (*Costruzione*). Ser Brunellesco lived from 1331 to a time between 1397 and 1404, see C. von Fabriczy, *Filippo Brunelleschi, sein Leben und seine Werke*, Stuttgart, 1892, pp. 1–4 (hereafter: *Fabriczy*); also *Cupola*, p. 202. We noted the recent appearance of Angela M. Romanini, *Arnolfo di Cambio e lo "stil-novo" del gotico italiano*, Milan, 1969, but have not seen this book.
[6] N. Machiavelli, *Istorie fiorentine*, 1532, II, viii; R. Davidsohn, *Geschichte von Florenz*, IV, 1, Berlin, 1897–1927, pp. 57 ff.; also see *Costruzione*, p. XXXIII. For approximate foundations of the old

centuries later, near the end of the feudal period, it was badly in need of reconstruction. In 1282, the city adopted a government of Major and Minor Guilds, and ten years later Arnolfo di Cambio, an architect of considerable repute in Rome and elsewhere, was hired to build foundations for a new front, new and wider walls, and new and higher piers, reaching to a level above the old basilica. It was also proposed—either then or perhaps during the next few generations, which included the times of Giotto—that the new cathedral would no longer terminate with small semicircular apses, as the original one, but with an enormous cupola structure.

The middle decades of the Trecento were one of the most miserable times that visited Florence and Europe, a time dominated by war, civil war, plague, famine, and insane movements. The Guild republics faltered badly. The Ghibelline Visconti and the Anjou, tyrants of Milan and Naples, respectively, were ascendant. Their political success had effects on Italian building styles. The Milanese court, full of Germans, as well as the southern court, usually full of French, attracted French and German followers, including Gothic builders, into Romanesque Italy. Their influx accelerated a movement that had begun already under Imperial and monastic influences in earlier times —the evolution of an international style containing many northern motifs and methods.

church, excavated after the flood of 1966, see our Figure 1 at "1." Detailed results have not been published yet.

1. Parts and dimensions of the Florentine Dome. Principal architects and their dates. Drawing of G. Rich, 1969, showing Cupola and Tribunes in top view, at left, and showing foundation in plan view, at right. *1.* The old cathedral, S. Reparata, built about A.D. 600. *2.* Nave and aisles of the new cathedral, S. Maria del Fiore. These parts designed and built by Arnolfo di Cambio, Francesco Talenti, and others, 1292–ca. 1375. *3.* "Giotto's Campanile," 1334–1359. *4.* The great Octagon, designed by Neri di Fioravante in competition with others, 1366–1367. *5.* Neri's Model, built by Ghini, 1367. *6.* Pier foundations and piers of the octagon, built by Ghini and others, 1367–ca. 1400. *7.* Tribunes and Tambour, built 1400–1418. *8.* Brunelleschi's model, built 1418–1420. *9.* Cupola of Brunelleschi, 1421–1432. *10.* Exedras of Brunelleschi. *11.* Lantern of Brunelleschi. *12.* Gallery proposed by Brunelleschi.

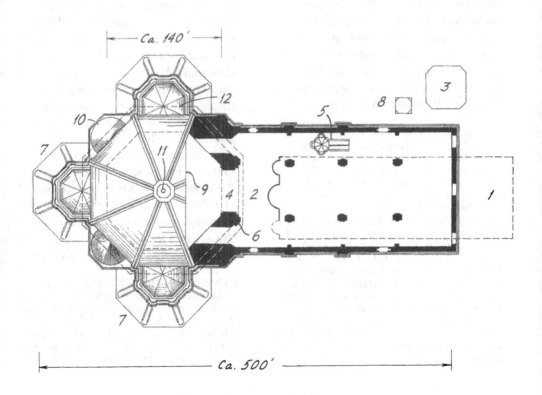

Major segments of the Florentine population hated the foreign influx with a passion. A movement took place that contained a variety of Guelph, Roman, and Humanist components; it may also be described more simply as an anti-Gothic or conservative movement.[7] In Trecento sculpture it brought strong renewals of classic Roman forms. In Trecento architecture it brought emphatic reassertions of Romanesque traditions.

It is unknown whether Arnolfo had planned a Gothic or a Romanesque form for a Florentine Cupola (see Figure 2). It is also unknown in part how he would have constructed its fabric and ornamented its surface. Some buildings in Gothic form had been built in Italy, but Italians were more familiar with the Romanesque traditions.[8] There were flat-profiled Roman cupolas, comprising an interior cap or *calotta*, hidden entirely or in part behind an outer wall. A totally hidden cupola was to be found in the Florentine Battistero. Whatever Ar-

nolfo may have planned, this was the kind of cupola known and revered in Florence during the Trecento.

Work on the Florentine cathedral was restarted after the Black Death, about 1350.[9] It still was unclear whether the cupola would be Gothic and pointed, or semicircular (Figure 3). However, while continuing to cover the walls with the usual geometric-polychrome, marble incrustation, the builders had begun to add Gothic windows and niches. The developing plans gave rise to power struggles. In 1361, the Woolen Guild announced that its Opera would no longer appoint officers or masters without consent of the Guild Consuls, because "workers must not be defrauded."[10]

Debates about the cathedral construction then dealt mainly with the reinforcements required by the large vaults, and their aesthetic effects. The *capomaestri*, Francesco Talenti and Giovanni Ghini, wanted a Gothic church with outer, reinforcing buttresses. Various advisers disliked this un-Roman kind of building and probably its Ghibelline associations. The position of some artists was complex. Orcagna in his so-called Tabernacle—the monument in Or San Michele, full of pictures, sculptures, and marble embroidery, reaching up to the ceiling—ornamented the top with a cupola on a tambour with circular windows. This

[7] These matters are very controversial, as the claims of the Ghibellines, Angevines, and others still find protagonists, while some schools of historic writing totally subordinate such claims to the asserted movements or "revolutions" in economy or technology or art. Figure 2, reconstructing the design of Arnolfo or his early followers, is from W. Paatz, *Werden und Wesen der Trecento-Architektur in Toskana*, Burg bei Mein, 1937, p. 84.
[8] H. Decker, *Gotik in Italien*, Vienna-Munich, 1964, has good illustrations of works throughout Italy, dating from Romanesque to Gothic, and showing the transitions from one style to the other (for example Pls. 14, 45, 56 ff., 131 ff.). We do not, however, agree with many of the views expressed in his text.

[9] *Costruzione*, Doc. 68 ff. and pp. LI ff.
[10] *Costruzione*, Doc. 84. Also see A. Grote, *Studien zur Geschichte der Opera di Santa Reparata zu Florenz im 14. Jahrhundert*, Munich, 1961, pp. 61 ff., D. Gioseffi, *Giotto architetto*, Milano, 1963, pp. 126–132, and literature cited. Details are controversial.

2. A possible reconstruction of the so-called Model of Arnolfo, ca. 1292–1302. After Paatz, 1937.

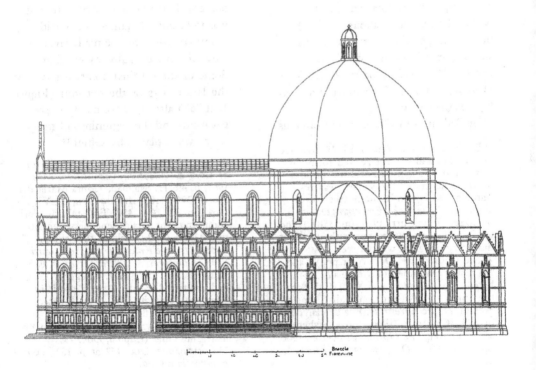

Braccia
5ª Fiorentine

feature alone is similar in outline to the now existing Dome, but an entirely different effect is produced by the surrounding four and higher pinnacles of Gothic form and ornamentation.[11] Even without the pinnacles Orcagna's dome form does not suggest technical possibilities for the construction of an unusually large vault. It shows one variant of Florentine-Gothic design, without an attempt to show Gothic or other structure.

The official debates of the time did not deal with the form or anatomy of a cupola structure. They dealt with the design of windows for the nave, which Ghini was finishing, and with proposed dimensions for the foundations of the cupola-supporting octagon at the end of the nave. Talenti had proposed that this octagon should have an inside width of 62 *braccia* (about 120 feet), but Ghini would have increased this size, apparently even beyond the present 72 *braccia* (about 140 feet). A chronicler wrote of a cupola almost as high as the actual one.[12]

For Talenti and Ghini it was obvious that the walls of the octagon would have the strong outer buttresses or counterforts of Gothic tradition. In keeping with such tradition, Ghini proposed that the walls themselves be made rather thin, thereby providing a maximum of church space within the inside boundaries of these walls. The outer boundary of the octagon was substantially given by the existing nave structure.

A different view was taken by a group of artists officially consulted by the Opera, who had as their leader the architect Neri di Fioravante.[13] Neri and his committee did not want buttresses. For this reason, and since the possibilities of using inner ribs were limited, these men needed stronger and thicker walls for their octagon. Their inside octagonal area became smaller than it was in Ghini's design, as the outside boundary could not be made larger in view of structures already existing. Some of the cardinal points for two of the large piers of the octagon (Figure 1, at "6") already were fixed by excavations and the beginnings of masonry work placed by Ghini.[14]

[11] Decker, *Gotik in Italien*, Pl. 75. Also see his Pls. 116 and 233 for related forms in Rome and Verona. The stylistic position of Orcagna, stated by Vasari, was restated by K. F. von Rumohr (see for example his *Italienische Forschungen*, ed. J. Schlosser, Frankfurt, 1920, pp. 358 f.) and by Rumohr's followers, but without major discovery as far as we can see.

[12] See *Costruzione*, Doc. 70 at pp. 81 and 131 ff. for the dimensions officially debated; Doc. 82 and pp. LXXVIII ff. for the chronicle of Marchionne di Coppo Stefani, *Istorie fiorentine*. For similar debates in Chartres, Milan, and elsewhere, see P. Frankl, *The Gothic*, Princeton, 1960, pp. 58 ff.; Holt, *Documentary History*, I, *passim*.

[13] About Neri, see H. Strauch, *A History of Civil Engineering*, Cambridge, Mass., 1964, p. 49; H. Saalman, "Santa Maria del Fiore 1294–1418," *Art Bulletin*, XLVI, 1964, pp. 477, 484, 489. To establish Neri's committee, the Opera sent messengers to two Major Guilds, asking for "the best," *quod placeat eis mittere sculptores aurifices et pictores . . . meliores quos habent* (*Costruzione*, Doc. 140). Allegedly Ghini had requested it, *ut . . . non possit in aliquo reprehendi* (Doc. 131). A descendant of Neri, called Nerius Francisci de Fioravantibus, was *operaio* in Filippo Brunelleschi's time (*Cupola*, Doc. 75).

[14] *Costruzione*, Doc. 152 at p. 182: *chome chominciate sono.*

Ghini asserted that the artists' plan would lead to a church with clear inside space smaller than his own. This was true. He also asserted that it made the structure weaker. This was very questionable.[15]

The artists consistently denied that their plan was less strong than Ghini's. At one point, as we will see, they charged that Ghini misrepresented their plan.[16] The negotiations were complex, although Ser Brunellesco, as a lawyer, may have understood them clearly.[17] So much is certain that Ghini vociferously insisted on the use of tall buttresses as outer reinforcements for the nave. He strongly suggested the use of similar buttresses for the cupola structure. He also wanted tall Gothic windows. The artists firmly rejected this design.

The citizens of Florence favored the artists' view. In 1364, they decided that "in the wall above the minor [side aisle] vaults there shall be round openings, not [tall, narrow] windows to admit light through the walls of the large [nave] vaults, one opening in each vault."[18] The existing "wall above the minor vaults" (Figure 11, directly below the roof of the nave) shows the actual use of the artists' ideas, while the lower outside walls show the design ideas of Arnolfo, Giotto, Talenti, and Ghini.

It hardly appears that the debates of the 1360s applied to any part of the octagon structure except its foundations, such as a drum or tambour for this structure. No tambour is mentioned in the documents of the time, although Orcagna—a member of the artists' committee—shows a tambour in his Tabernacle, as mentioned. The debates dealt with the superstructure of the nave and the substructure of the octagon, and the decisions had to do with these structures exclusively, in keeping with the empiric step-by-step methods used by early builders.

Friction between Ghini and the artists was very evident in the summer of 1366, when Ghini was constructing the first two vaults of the new nave, near the entrance facade. There were no buttresses leaning against the vaults, and the old-timers were nervous when Ghini placed the masonry on his conventional centerings. They became very nervous when hairline cracks appeared (which unavoidably result from the shrinkage of the cement during the setting of the masonry).[19] The artists

15 The statements of Ghini are reported for example in *Costruzione*, Docs. 150, 174 (August 1366 and July 1367). They are vague as to dimensions. He only says his choir structure is smaller outside and yet larger inside than that of the artists (*occupa meno terreno per largheza ed e capace di piu gente*). This is conceivable if his walls are thin.

16 *Costruzione*, Docs. 176, 178 (31 July and 9 August 1367).

17 The matter becomes still more complex by accounts of historians who apply arbitrary selections to the surviving evidence. Such an account is given by P. Frankl, *Gothic Architecture*, Baltimore, 1962, who simply disregards Ghini and Neri. In his view the ultimate Cupola is "the concrete expression of Talenti's ideas," and the presence of laymen in the Opera supposedly "produced the spiritual atmosphere" that enabled Brunelleschi, supposedly, "to break with the traditions of the Gothic style" (p. 224).

18 *Costruzione*, Docs. 119 ff. (end of 1364).

19 *Costruzione*, Doc. 143: *apparivano certi peli.* For the other events mentioned here, see Docs. 140, 144, 149.

were undismayed. They enforced exclusive reliance on tie rods, so-called chains.[20] They also established themselves as a permanent supervisory committee. Tie rods as specified by the artists extend across the arches between the nave and the aisles. The rods are ugly as they cross the arches in odd locations.

As to tie members or other reinforcements for the octagon, the artists were hardly expected to have finished proposals, and they expressed only general ideas about this problem. In a report dated August 1366 they expressed their hope that someone would find some means to construct the cupola without visible chains or ties.[21] Their report is silent about a tambour and about the height of the cupola. The artists' design was recorded on parchment sheets, as were the counterdesigns of Ghini and others.[22]

For some time the outcome of these proposals was in balance,[23] but dramatic proceedings that followed brought an end to the era of strong, Gothic influence. A civic committee decided in favor of the artists, and the Opera then ordered construction of a masonry model based on the artists'

design.[24] Ghini, as *capomaestro*, constructed it and also continued to build the nave of the church.[25] In May 1367, Ghini obtained a hearing about alleged dangers for the cupola, inherent in the artists' design. A committee of other artists convened, which proceeded to express doubt in the structural soundness of Neri's plan. In July the questions came before a new body of civic advisers. It was then that Neri complained about Ghini's misrepresentation of the artists' design.[26] Ghini apparently had shown the inside area very large, while leaving the outside boundary as required by the nave, and had omitted outer reinforcements as requested by the artists. Thus he had weakened the piers while adding to the vault pressures.

The new civic committee decided that the danger alleged by Ghini could be eliminated and that he should return to the basic inside dimensions, which the committee now specified in very emphatic terms.[27] Ghini was to use them in producing a corrected model of the artists' design. He also was given a substantial loan. It seems that Ghini then relented. He built the corrected model, as well as the end of the nave, and there is no indication of further friction. The artists made a concession of their own: they allowed the use of triangular buttresses reinforcing the tribunes.[28]

[20] *Cupola*, Doc. 383 (1369).
[21] *Costruzione*, Doc. 153: *Senza esservi catene che si veghano.* The same hope was entertained, later, for the vaults in the Dome of Milan and in SS. Giovanni e Paolo in Venice, but the actual structures were disappointing (see Decker, *Gotik in Italien*, Pl. 7 and p. 205).
[22] *Costruzione*, Docs. 150, 300, 341, and p. XCVII. Initially, the artists provided only a plan view. Ghini remarked on the lack of a profile view, *disegno dell' altecza* (Doc. 150).
[23] *Costruzione*, Docs. 146, 147.

[24] *Costruzione*, Docs. 150, 153, 155, and pp. LXXXVIII f.
[25] *Costruzione*, Docs. 158–163, 189.
[26] *Costruzione*, Docs. 170, 174–176.
[27] *Costruzione*, Docs. 178, 181, 271.
[28] A. Nardini-Despotti-Mospignotti, *Filippo Brunelleschi e la cupola del Duomo*

Ghini's model was then destroyed, while the artists' model, built by him, was definitely adopted.[29] This model then remained in the aisle of the Dome near the Campanile.[30] For some fifty years it stood near the area where the visitor now stands to see the monument to Brunelleschi. (See our frontispiece. The area at right of "5" in Figure 1 is its location.)

In October 1367, the city organized a large, new committee to make a final decision in the dispute between Gothic and Romanesque factions.[31] The committee included several dozen members of Major and Minor Guilds who were selected, as usual, by lot from a hand-picked list. One member was Ser Brunellesco, who had just returned from a trip abroad as a lawyer in the service of Florence. Other noteworthy participants were Antonio Machiavelli and Jacopo degli Alberti, who may have been merchants. Jacopo stated that he found the artists' model more beautiful outside—no doubt because it omitted the Gothic appendages—and that Ghini's model seemed more beautiful inside, probably on the ground that Ghini more definitely than the artists omitted visible tie rods, spanning the vault in various directions. The committee expressed itself in favor of the artists.[32] Then, as a final step, came a

referendum of some five hundred citizens. It confirmed the conclusion, as referenda in Florence usually did.[33]

Neri's committee had obtained this decision by being adamant in its rejection of Gothic vault structure, although some of the artists had sympathetic attitudes toward Gothic ornamentation.[34] The committee had the support of numerous and influential Florentines. It was powerful enough to obtain disciplinary action, at various times, against masters who favored Gothic principles.[35]

The committee's victory probably led to feelings of national and aesthetic satisfaction. However, the artists and their model left a large problem unsolved, which Filippo then inherited. This was the problem, how to overcome the dangers anticipated by Ghini, while omitting his buttresses. No one at the time, or for centuries thereafter, could solve such a problem theoretically.[36] The record indicates that the artists did not even attempt to solve it. Their decision, like Arnolfo's original design, was aesthetically bold but structurally weak, and its weakness did not go unnoticed in later times.

EARLY STUDIES

The tangible result of the great debate was the artists' model, built by

di Firenze, Florence, 1885, p. 125 n. 2 (hereafter: Nardini).
[29] Costruzione, Docs. 155, 189, 192, 193. Destruction of the old Santa Reparata occurred some ten years later (Doc. 237).
[30] Costruzione, Doc. 189.
[31] Costruzione, Docs. 185, 214.
[32] Costruzione, Doc. 190 at p. 202 and p. XCVI.

[33] Costruzione, Doc. 190 at pp. 200–205.
[34] Costruzione, Doc. 70 at p. 100 and p. LXIX.
[35] For example, Costruzione, Doc. 126 (1364).
[36] Strauch, A History of Civil Engineering, pp. 51–162.

Ghini. It was later called a small model, by comparison with another structure. Probably it showed the then-existing building portions and the more or less clearly proposed additions on a scale of no more than about 1:16, that is, with little detail. Perhaps one could just barely enter the nave and see the interior. The documents do not indicate whether the model had any kind of cupola or tambour. They only state that it had three tribunes of five chapels each, and "certain walls" in their vicinity, probably meaning the buttresses contained in the compromise plan.[37] The debate had made it clear, and the model probably confirmed, that with minor exception neither the superstructure of the nave nor the understructure of the octagon was to be Gothic.[38]

This model was one of the prize possessions of Florence during the decades about 1400. Filippo must have seen it often, and since his father was able to describe its background history, Filippo soon must have learned how much the model left unanswered.

He may also have seen a painting of the then-proposed cupola and church, painted by Andrea Bonajuti da Firenze about the time when Ghini built his various versions of the artists' model.

Andrea's painting, like the artists' model, was revised in its representation of the cupola structure.[39] Between the two half domes covering the tribunes, pinpricks indicating the sloping line of a large buttress are still visible in the plaster. Also visible is the ornamentation originally painted along this sloping line. Thus it appears that the painting was planned and begun in Ghini's form but was changed in response to the artists' view (and Andrea belonged to a subcommittee of the artists' group). Shadowy reflections of the change can be seen in good photographs of the painting (Figure 3, see detail view II). To a sharp-eyed viewer, such as Ser Brunellesco's son, it could become apparent that the artists' opinion had been controversial and that it still was problematic.

Large buttresses supporting a cupola are also shown in other pictures in or near Florence[40] and are used in other churches.[41] However, could this kind of structure prevail in Florence after 1367? And if not, what kind of structure constituted a safe support for the

---

[37] *Costruzione*, Docs. 207, 248.

[38] A fundamentally different view is stated by *Nardini*, p. 46 and *passim*, and by his followers. However, the only factual evidence, adduced by Nardini, is that in 1426 someone spoke of "the contour decided upon . . . several decades ago." This expression fits the year 1407 as well as the year 1367; so also Guasti, in his introduction to *Costruzione*, p. CXIII.

[39] *Costruzione*, Doc. 155 (1366) shows the date and the specific connection with the Florentine Cupola. The changes were ascertained in 1948 through the courteous and able assistance of Dr. G. Marchini.

[40] A close variant of such a structure appears in the Lorenzetti fresco in Assisi, *Christ's Entry into Jerusalem* (Decker, *op. cit.*, Pl. 90). There are variants of more abstract character in Giotto's *Massacre of the Innocents*, Padua, and in Taddeo Gaddi's *Meeting at the Golden Gate* in Santa Croce, Florence.

[41] For example, San Francesco, Bologna (Decker, *op. cit.*, Pl. 187), where the hidden half dome is supported by flying buttresses between outer chapels.

immense cupola to be built? Ser Brunellesco could not answer such questions, but he could prepare the receptive mind of his son. The various pictures of cupolas in public places—probably more of them, at the time, than we know now—could provide the child with images to consider, to adopt, or to reject.

The biographers overlooked these early influences. They held Trecento architects in contempt. They used stereotype, describing Ser Brunellesco as a resigned father, who merely passively permitted his son to learn the arts of jewelry or sculpture.[42] The documents show that, more probably, it was the father who directed Filippo's thought to the Cupola plans, their administrative history, and their technical problems. During the years about 1390, when Ser Brunellesco was aging and Filippo was adolescent, the Opera began to raise the massive, unbuttressed octagon walls designed by the artists, and about this time Filippo's mind must have received an original germ of its vast conception. Nothing similar to the final structure was yet to be seen on the ground (Figure 4), and only some few of the decisions needed for the Dome had been made, no matter how often the Florentines insisted on compliance with the artists' model. New thought was needed. It appears that Filippo, at this early stage of his life, had already become a critical reviewer of the builders' progress. This became obvious before long.

During the later 1390s, the Opera raised the two end piers of the octagon —massive structures which now contain the sacristies.[43] While this work proceeded, there were further debates in the building yards. Old-timers questioned whether the piers designed by the artists were strong enough "for the great load to be placed thereon." New committees and administrators repeated the affirmative answers of the artists' committee, but did not, so far as the record shows, make these answers more tangible or plausible.[44] Young Filippo must have learned of the new debates. It seems he responded by taking a historic decision. Manetti plausibly reports that the young man, while taking some interest in sculpture, jewelry and clock-work,[45] decided "to rediscover the manner in which the Ancients had built." Vasari repeats the statement and refers specifically to young Filippo's interest in the cupola problem.[46]

About 1400, the piers for the support of a cupola were complete so far as any of the builders then knew. The arches connecting the last piers were being constructed;[47] and provision was made for the possible use of tie rods, crossing these arches. No doubt the people con-

[42] Manetti, p. 6: 'l padre gliel consenti, che era huomo prudente.

[43] Costruzione, Docs. 302, 309, 312, 410. In Filippo's time these piers were provided with elevated singers' porches, cantorie, which later were converted into organ lofts.
[44] Costruzione, Docs. 256, 352.
[45] F. D. Prager, "Brunelleschi's Clock?" Physis, X, 1968, pp. 203–216.
[46] Manetti, pp. 18–23; Vasari, p. 202. Both writers suggest, but do not state, that the decision came after the Battistero competition of 1401. We think the decision was at least partly formed in Filippo's earlier youth.
[47] Costruzione, Docs. 411–418.

3. The Cupola as foreseen by Andrea Bonajuti di Firenze, ca. 1367.
I. Detail from *The Militant Church*, fresco painting in the Spanish Chapel of S. Maria Novella, Florence. After an old photograph.
II. Enlarged portion, emphasizing the outline of a buttress, which is dimly visible also in unretouched picture *I.*

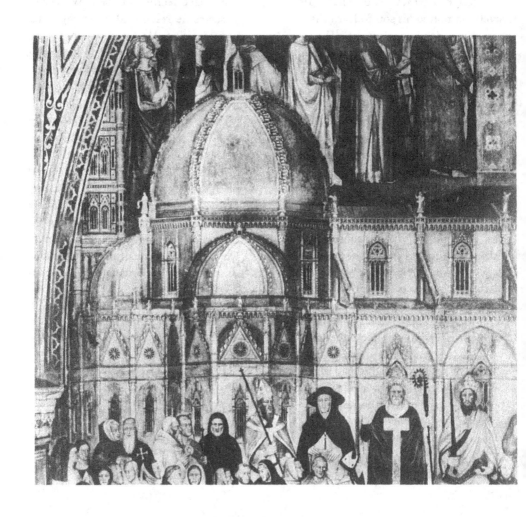

4. The Dome in 1390. Schematic reconstruction of its appearance seen by young Filippo Brunelleschi. Drawn by G. Rich, 1969. *1.* The Romanesque facade and sidewall begun by Arnolfo, partly converted into Gothic form. *2.* Talenti's Gallery. *3.* The circular windows designed by Neri di Fioravante and built by Ghini. *4, 5.* First and second piers, designed by Neri and built by Ghini. *6.* First major arch, built by Ghini. *7.* Excavations for the foundations of the Sacristy piers and the Tribunes.

tinued to discuss the construction. This also was the time directly before the famous competition for the Battistero doors, and according to popular account that competition had something to do with Filippo's becoming an architect.[48] In our view he already was well on the way to becoming an architect of uncommon creative power. In his well-prepared mind, the question was bound to arise whether it would be necessary to use the tie rods across the arches and also to use such rods in the cupola itself. As noted by the biographers, he asked himself whether the Ancients knew better. His actual study of the Ancients, in the ruins of Rome, was promoted by the fact that he failed to be recognized as the first of the Florentine sculptors then working. But we can hardly think this failure was the only reason, or the main reason, for his going to Rome.

Others reacted very differently to the building problems of the time. A new architect, Giovanni d'Ambrogio, came into office about 1400; and he was an outspoken old-timer who favored Gothic form.[49] He showed this promptly in his first major undertaking, the completion of basic structures belonging to the first or eastern Tribune of the Dome, opposite the nave. The walls of the five chapels, surrounding this Tribune, had been built and vaulted during the last years of the Trecento. In 1403, Giovanni acquired timber for the flat roofs above the vaults of these chapels.[50] He noted that these roofs would make it impossible to see most of the Tribune structure itself, and of the windows in its walls, unless the structure and its windows were raised to a level higher than a preestablished gallery, extending around the entire cathedral.[51] Giovanni proceeded to raise the upper ends of these windows, and of the buttresses between them, although they then interrupted the gallery (Figure 5).

The reaction was prompt and sharp. A new board of nineteen advisers, notably including Ghiberti and Brunelleschi, decided in 1404[52]

[48] The story of the competition is well known from *Manetti*, pp. 14–18 and *Vasari*, pp. 200 f.
[49] Giovanni, together with his father Ambrogio, had questioned the structural sufficiency of the artists' model in 1384 (*Costruzione*, Doc. 352), and, as already noted, there was good reason to question it. A brief biography is given by G. Brunetti, "Giovanni d'Ambrogio," *Rivista d'arte*, XIV, 1932, pp. 1–22.

[50] *Costruzione*, Doc. 423. Photographs of these roofs and surrounding regions may be seen in P. Sanpaolesi, *La Cupola del Brunelleschi* (Forma e colore. I grandi cicli dell' arte), Florence, 1965, Pls. 11, 29 (hereafter: *Forma e colore*). See also our Figure 5.
[51] As a result of Giovanni's argument, the Opera exhibited old drawings (probably of 1367) at the Campanile (indicating that the model of 1367 did not show the details in question or at any rate did not show them clearly enough).
[52] *Considerato quod super constructione . . . spronorum ad praesens edificatorum per Johannem Ambroxii c.m. fuit commissus certus error . . . deliberaverunt: che lo sprone mosso per Giovanne d'Ambrogio fuori delle debite e vere misure si rivochino e amendino in questa forma, cioe: che no[n] rimutando il piano dove i detti sproni si principiono dicesi che il detto sprone si deba abassare e fare che lo sprone venga a finire a quella parte di fuori proprio chome e il disegnio ch'e al campanile, con questa che dagli sproni in su si faccia l'andito co' beccatelli e col para-*

. . . that a certain error was committed and that the buttress begun by Giovanni d'Ambrogio at variance with the required and true measures must be undone and amended so that, without change to the plane where these buttresses begin, this buttress must be lowered so that the buttress will finish at such part of the outside [wall] as is shown by the design at the Campanile, so that the gallery with parapet on brackets will be above it.

Perhaps in order to soften the blow, the advisers agreed with Giovanni that according to their decree insufficient room remained for desired window ornamentations and that "four masters shall consider with Giovanni d'Ambrogio what is better: to lower the window opening or to leave it as it is and shorten or suppress the window frame."[53]

Giovanni obeyed and lowered the buttress.[54] Perhaps it would have been

5. Dome designs of 1404. Schematic reconstruction, drawn by G. Rich, 1969. *I.* Design proposed by Giovanni d'Ambrogio. *II.* Changes enforced by the committee of 1404, including Ghiberti and Brunelleschi. *III.* The change of 1409, adding the Tambour. *1.* Tall buttresses proposed and started by Giovanni d'Ambrogio, in contrast to *2*, the low buttresses actually used and existing. *3.* Tall windows proposed by Giovanni d'Ambrogio, allowing better sight of the windows from the street level, compared with *4.* the Tribune windows actually built, which are visible only in small part from the street. Comparison of *1* and *3* with *2* and *4* shows that Giovanni would have built a Dome of conventional Gothic form from bottom to top and that the committee of 1404 kept the use of such form limited to the understructure, as the committee of 1367 had proposed. *5.* Gallery of Talenti. *6.* Tambour. *7.* Area of the Gallery subsequently proposed by Brunelleschi. *8.* Exedra. *9.* (Hypothetical) One of a set of tie rods, strengthening four of the eight Tambour walls.

---

petto. The text alternates between *sprone* and *sproni,* as Giovanni had just begun the first of his twelve buttresses. That the text also shows irregularity, such as the phrase *lo sprone . . . si rivochino,* may be taken as a sign that the debate was agitated.
[53] The documents have been mentioned but not interpreted in the literature. For example, R. Krautheimer, *Lorenzo Ghiberti,* Princeton, 1956, pp. 4 f., 254 f., disregards the fact that special reference was made to Ghiberti in 1407. P. Sanpaolesi, *Brunelleschi,* Milan, 1962, disregards the entire episode. This latter book (hereafter: *Sanpaolesi*) shares the view of the modern literature (perhaps traceable to Guasti and his dichotomy between *Costruzione* and *Cupola*) that Brunelleschi's architectural work began in 1417. We think the view of Manetti and Vasari is sounder and in better agreement with the documents.
[54] Also see *Costruzione,* Docs. 433 ff.

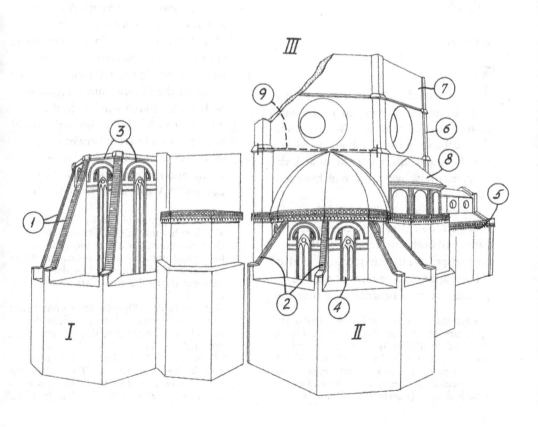

even better if the buttresses had been abolished almost entirely, but although they were a mere compromise, they had to stay because they were shown by the artists' model. The Opera retained them, just as it had retained the Gothic windows of the understructure when it provided for the Romanesque windows of the upper part of the nave. The artists of the earlier generation had vetoed the use of buttresses outside the nave, and the new artists now vetoed the use of buttresses along the spring line of the cupola. The debate caused administrative action adverse to Giovanni, although not permanently.[55] The Opera closed the affair with a minimum of agitation. It did not close Filippo's mind.

THE TAMBOUR

In the recorded debates of 1404–1405 there is still no reference to a Tambour or to Cupola details, but the step from these debates to the Tambour decision is a short and direct one, or so at least it appears now when the step has been taken.

The decision to build the Tambour followed soon after these earlier debates, and both matters dealt with applied perspective. It seems to us that Filippo attacked questions of applied perspective on a broader scale than the

earlier literature has noted. He dealt with such questions many times, for example when he studied the perspective appearance of the Battistero, from inside the portals of the Dome; again when he undertook the development of statues to be seen from below the buttresses of the Tribunes; also when he acted on the committee of 1404–1405; and also, according to one important report, at the time of the Tambour decision, made a few years later.[56]

It does not follow that Filippo applied very exact methods of perspective. His procedure was likely to be empirical, even if some friend explained to him what early scientists had written about rays of vision.[57] The problem was how to build and how to paint. About these matters early science was silent.

Vasari writes that Filippo, a few years after the Battistero competition, gave advice to the Opera del Duomo in favor of lifting the Cupola onto a Tambour with a large round window in the middle of each side "for not only would it lift the weight from between the shoulders of the Tribunes but it would also make it easier to vault the Cupola."[58] Vasari adds this to the ac-

[55] *Costruzione*, Docs. 435, 439: temporary appointment of a second architect, Guidi, followed soon after by his removal, and ultimately followed by dissolution of the committee. Ghiberti, as a committee member, received 3 florins, which he had to return in 1407 (*Costruzione*, Doc. 438).

[56] *Manetti*, pp. 10–13; *Costruzione*, Doc. 438. The interest in visibility had caused the Florentines, as early as the 1330s, to lower and remove the then-existing structures near Corso Adimari in order to open the sight to the Duomo (*Costruzione*, Doc. 53).

[57] P. Sanpaolesi, "Ipotesi sulle conoscenze matematiche, meccaniche, e statiche del Brunelleschi," *Belle arti*, II, 1951, pp. 25 ff., especially 32, 35.

[58] *Vasari*, p. 203. We have slightly modified De Vere's translation. The Italian text reads: *Intervenne Filippo e dette consiglio ch' era necessario cavare l'edificio fuori del*

count that he takes from Manetti. Neither mentions the consultations of 1404–1405, but it is possible that Vasari interprets the transactions of 1404 to ca. 1410 as a single unit and that he treats these transactions as significant only insofar as they have to do with the Tambour.[59]

Modern writers frequently differ with Vasari, but not always on grounds of evidence better than his report. The documents show that late in 1409 the Opera invited the public to give advice about "the work to be done or being done."[60] Evidently a new proposal was in the air, and everything indicates that it was the Tambour proposal. Early in 1410, the Opera decreed that "construction of the [structure containing the] round windows of the great Tribune be executed and that their masonry be built promptly."[61] This is the first explicit reference to "round windows of the great Tribune," that is, to the Tambour.

Critics of Vasari's statement can cite a variety of early tambours and tambour pictures, in support of their premise that denies Brunelleschi's role in the project. However, there is a vast difference between the general concept of a tambour-cupola combination (various forms of which were old) and the particular Tambour-Cupola combination of Florence (which may still be a new and more specific creation of the time about 1409). This latter combination differs as widely from earlier tambour-cupolas as the famous Arcade of Filippo's *Innocenti* porch differs from former Roman and Romanesque arches, such as Orcagna's *Loggia dei Lanzi*.[62]

The real question is whether any evidence shows that either Filippo or others progressed from the general concept of some cupola-tambour to the particular form used in Santa Maria del Fiore. Guasti, a most competent historian, at least once declared it possible that Filippo took this step, as stated by Vasari. The ingenious and persuasive Nardini, on the contrary, announced that the step had been taken in the Trecento, perhaps by Ghini or possibly by Arnolfo.[63] Orcagna is also men-

---

tetto . . . fare un fregio . . . e in mezzo a ogni faccia fare un occhio grande, perche, oltre che leverebbe il peso fuori delle spalle delle tribune, verrebbe la cupola a voltarsi piu facilmente. Until *Nardini,* the literature followed the view of Vasari.

[59] Also see the technically interesting report of G. B. Nelli in *Cupola,* Doc. 391, and the more detailed version of Filippo Baldinucci (ca. 1650) in his *Vita di Filippo di Ser Brunellescho,* ed. D. Moreni, Florence, 1812, p. 188. Baldinucci (pp. 158 ff.) criticizes many views stated by Vasari, but he refrains from doing so at the present point.

[60] *Costruzione,* Doc. 453. *Manetti,* p. 30, says Filippo was then in Florence.

[61] *Costruzione,* Doc. 454. Surrounding circumstances are not reported in the published documents; the publication of facts for the years 1406–1418 is unsatisfactory.

[62] Most of the structures and pertinent pictures, including several by Giotto and his school, show hidden cupolas, surrounded by either low walls or wall segments (as in the Cathedral model of Pavia, reproduced in *Sanpaolesi,* Pl. 63) or by the pinnacles of "lantern towers," as in Milan and Fossanova (see Decker, *Gotik in Italien,* Pls. 2 f., 124 f.; also see C. Ward, *Medieval Church Vaulting,* Princeton, 1915, pp. 33, 119).

[63] As to Guasti, see *Cupola,* pp. 189 f., and *Costruzione,* pp. LXXIX f.; on the other hand, *Nardini,* pp. 97, 157, followed for

tioned. However, no evidence indicates that anybody, prior to ca. 1404, developed the actual Florentine Tambour design. Certainly the Opera had not adopted it.

During Filippo's period of activity, by contrast, the Opera adopted a Tambour plan and carried it out at once through Giovanni d'Ambrogio. It did this about the time of Filippo's perspective studies.[64] It had the Tambour built with smoothly outlined corner pilasters, foreshadowing the white marble ridges of Filippo's cupola. It may also have used horizontal reinforcements, along the upper and lower edges of the Tambour, in a new arrangement. There is evidence in the settling pattern, shown by certain cracks in the structure, that strong ties are inserted across and above the large Gothic arches, in four sides of the great octagon. Against every normal expectation these four sides are free of vertical cracks, existing in the other four sides. Some one must have put strong horizontal or diagonal reinforcements (tension rods) in the four weaker parts of the structure, converting them to stronger parts.[65] He must have arranged thereby,

in the massive Tambour masonry, to "make it easier to vault the cupola" (safely). The artists of 1367 had only raised a pertinent question. Giovanni d'Ambrogio and others are not known to have made pertinent proposals. We think Vasari's report that Filippo proposed it is plausible.

The reinforcements of the Tambour, in turn, made it possible to raise the Florentine Cupola in new and impressive form, very different from the flat or sunken forms of cupola used in Ancient and Romanesque times and also very different from the slender, towering, embroidered variants of Gothic *tiburios*. The new form expresses total rejection of the compromises for the understructure that had remained in the model of 1367. The debate of 1404 had lowered the compromise buttresses—now the cupola support was raised and strengthened. The new structure was totally separated from the old, so far as conditions allowed. The builder of 1410 decided to absorb the Cupola thrust in the Cupola, to counter other forces by ties in the Tambour, and to make the resulting structure freely visible over the underlying, medieval mass.[66]

It seems most improbable that such

---

example by C. von Stegmann and H. von Geymüller, *Die Architektur der Renaissance in Toskana*, I, Munich, 1885–1893, pp. 38 f.
[64] Also see *Fabriczy*, pp. 45–51; L. H. Heydenreich, "Spatwerke Brunelleschis," *Jahrbuch der preussischen Kunstsammlungen*, 1931, pp. 14 f.
[65] We owe this recognition to a private communication of Henry Millon. A hypothetical indication of such reinforcements is shown in Figure 5 at "10." The form of the Tambour wall, without marble incrustation, may be seen in the picture of

Dante, the Divine Comedy and Florence by Domenico di Michelino, in the Cathedral (P. Sanpaolesi and M. Bucci, *Duomo e Battistero*, Florence, 1966, Pl. 1).
[66] *Contra*: Saalman, "Santa Maria del Fiore," p. 486. He writes that only one major element, a Tambour proposed by Orcagna and Neri, could account for the Florentines' finding of greater beauty in the artists' project. It seems to us that the omission of Gothic buttresses and pinnacles could also account for the Florentine view.

decisions were taken by Giovanni d'Ambrogio. Of him we hear only that he relied on outer buttresses, even beyond the extent that the model of the artists provided them. Who but Ser Brunellesco's son was likely to impose his will on the master?

Documents for this time are rare, but one of them at least shows that Filippo was in contact with the Opera, as he had been in 1404. The document shows that Filippo received 10 soldi (less than a sixth of a florin) for brick, about the time when the Tambour construction began.[67] The payment could not reimburse him for much more than a bushel of this commonplace material. This is an odd transaction, which has not been explained. Did Filippo deliver model bricks to the Opera? If so, were they for the Tambour?[68]

According to Vasari's report, which we have cited, the Tambour was to make the vaulting of the Cupola easier, and the structure indicates that the Tambour at least made safe vaulting easier. Admittedly the statements are not explicit: the biographer's report is laconic, and the Opera documents are trivial. Therefore, interpreters of course may disagree. Yet there is at least some evidence indicating that Filippo about 1409 or 1407, if not already in 1404,

proposed to build the Tambour and Cupola in structurally and aesthetically new forms, different from both Gothic and Roman ones although containing elements of both. The new structure is simple and timeless and is very superior to the plans promoted by Giovanni d'Ambrogio, which would have relied on buttresses and inner or outer half domes for stabilizing the Cupola. The half domes were rather worthless, as the pressures existing in the octagonal Cupola necessarily are channeled into its corner ribs. For absorbing that pressure, it is useless to provide a half dome the top of which leans against the middle of a side. Therefore it was a valid and progressive idea to reject the buttress compromises, to absorb all forces acting on Tambour and Cupola in their own structure (except their weight), and to raise their fabric far above the understructure.

Whether Vasari's report had a documentary basis, beyond the facts cited here, is unknown. The report is not otherwise confirmed by documents now known.[69] Supplying bricks and conducting perspective experiments has no close relation to a Tambour decision; yet these activities establish at least some little support for Vasari's story

[67] 31 January 1410–1411: *A di 30 di settembre 1410 a Pippo di Ser Brunellesco per uno staio di mattoni presta soldi 10.* This document was published by K. Frey, *Le vite di Filippo Brunelleschi* etc., Berlin, 1887, p. 155.

[68] "Token" payments were also usual in the Opera's relationship with others: *ensenium extimationis . . . ut est in similibus usitatum* (*Cupola*, Doc. 1).

[69] We regret that we cannot now review the Opera documents in relation to research on Filippo's perspective work as published by A. Parronchi and others. As a result of the flood of 1966, the documents of the Opera are unavailable. Was Brunelleschi's construction of perspective views accompanied by the use of geometrically constructed orthogonal views? See our notes about Figures 8 (attributed by Nelli to Brunelleschi), 31 (by Giuliano da San Gallo) and 35 (attributed to Neroni), pp. 28, 101.

about Brunelleschi's inception of the Tambour, while the opposed view of Nardini, that it was conceived much earlier, seems to have no support at all if the Florentine Tambour decision is distinguished from the structurally vague tambour ideas previously debated.

BRUNELLESCHI'S
"MASONRY WORK"

2 AFTER THE TAMBOUR

The biographers indicate that Filippo
was often asked for his architectural
opinions, around the time of the Tambour construction.[1] About this same
time, Filippo also fascinated the
Florentines by his disclosure of an
unexpected technical idea: vaulting the
Cupola without an armature. Until then
the citizens had discussed the use or
nonuse of permanent reinforcements
for the vault, such as buttresses or
tie rods, but it appears that they assumed without question that the vault
(or at least the complete system of
vault-supporting ribs) would rest on
centerings held by armatures, until the
structure was completed to the top and
keystone. Filippo now proposed to
omit such centering armatures as well
as the Gothic buttresses.

The exact origins of his idea are not
reported. As noted, he may have
planned vaulting without armature as
early as 1409. A pertinent event occurred in 1415 when the Tambour was
two years old and when one of the side
Tribunes (the northern one) was ap-
proaching completion.[2] About this time

it became imperative to make the decision that the builders had dreaded for
many years, that is, to design the inner
and outer wood structures, believed to
be needed for support and hoisting of
the materials. It had been easy to raise
the Tambour materials, by some plain
hoist inside the octagon, but now it
seemed necessary to close the octagon
(Figure 6) and to build complex wood
structures both inside and outside
(Figure 7 at XII and XIII). How were
these enormous scaffolds to be built?
Where was the lumber for them to be
found? How could they be made
sufficiently rigid?[3]

Immediate answers to such questions
were not forthcoming, but there were
rumblings about current sculptural
work. Donatello and his father, Nic-
colò di Betto Bardi, had for some time
designed statues for the Dome, and
Filippo had joined them in this enterprise. In or before 1415, they had contracted to provide "a small statue of
stone, with gilt cover of lead, to be
made as a model and demonstration of
the large statues to be placed on the
buttresses" of the understructure, in the
area of the Tribunes. An advance of 10
florins was authorized on 5 October
1415. At this point Filippo disappeared,
as he did repeatedly during his early
years. Vasari says, "Being taken one
morning with the idea of returning to
Rome, he went there, thinking that he

---

[1] *Manetti*, p. 30. Lorenzo and Filippo may
well have been present on 21 August 1414
when a new Council, apparently in the absence of Giovanni d'Ambrogio, once more
decided to lower the top of a tribune, this
time by only 8 inches. Advice about it had
been given by "many" (*da più e più persone*), *Costruzione*, Doc. 421.
[2] *Costruzione*, Doc. 454. The three Tribunes were built, respectively, in 1395–
1406, 1407–1415, and 1416–1421, or in these

approximate years, as indicated by *Costruzione*, Docs. 401, 410, 440 ff., 484 ff., in
addition to the pertinent documents already cited.
[3] Such questions were recognized; see
Costruzione, Docs. 446, 447, and 486.

would be in greater repute and would be more sought for from abroad." He did not succeed at once. The Opera, on 16 January 1416, threatened him with jail for failure to comply with his contract.[4]

A year later this excitement had subsided in part. The statues for the buttresses seemed forgotten for the moment. (Later, more of them were made. The marble block from which the *David* by Michelangelo was carved was originally planned as one of them.) In 1417, Filippo was engaged "in making designs and in labors for the Opera in Cupola matters." It may have been then that he produced a sketch, pertinent to the idea of vaulting without armature, a copy of which was later attributed to him (Figure 8). It combines the inside profile of a typical Roman or Renaissance cupola with the proposal to erect a scaffold, not a centering, in the cupola space.[5] The drawing does not show when it was that he first proposed to vault the Cupola of Santa Maria del Fiore with the aid of such a scaffold; it only shows that he (if he really produced the sketch) planned to use vaulting without armature in a variety of cupolas.

6. The Dome and its armature problem, 1413–1418. Drawn by G. Rich, 1969. Buttress *1* had been lowered to the level shown here, while Tambour *2* had been built, and Side Tribunes *3* had been begun. The next question was how to build armature *4* for the support of vault-spanning centering *5*. No one had yet proposed to vault without armature (Figures 7, 8) or to use the ultimate, cellular Cupola structure with internal buttresses and walks (Figure 9) or its tie rings (Figure 10), but some visible features of the new concepts began to appear, for example, element "7" of 1404 (Figure 5). However, the builders continued to worry about the strength of Sacristy piers, *6*.

[4] *Costruzione*, Docs. 472, 473, 475; C. von Fabriczy, *Brunelleschiana*, Berlin, 1907, pp. 10 ff. (hereafter: *Brunelleschiana*). As to the statues originally planned, see our Figure 3. As to those installed in the sixteenth century, see the title page of the book by Sanpaolesi and Bucci, *Duomo e Battistero*.
[5] See for example Turin, Cod. Saluzziano 148, 84^r-d, 84^v-a, Francesco di Giorgio, *Trattati*, ed. C. Maltese, Milan, 1967, Pls. 155 f. (hereafter: *Trattati*). Also see the well-known inside profiles of the Pantheon in Rome and Santo Spirito or the Pazzi Chapel in Florence.

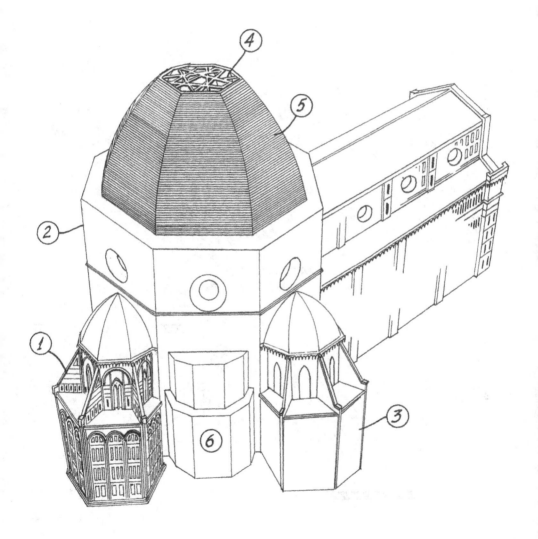

7. Vaulting with and without armature. Schematic drawing by G. Rich, 1969. *I.* With armature. *II.* Without armature. *III–X:* simplified indication of a pattern of bricks or stone blocks that can be placed with or without armature; not intended to show the pattern used in Brunelleschi's Cupola, which is largely unknown. If a conventional centering, *XI,* is used, this requires inner armature, *XII,* and outer scaffold, *XIII;* if not, the work only requires work platforms, *XIV,* which can be attached to a relatively small and light inner scaffold, *XV.* The masons can still use portable centerings, *XVI.*

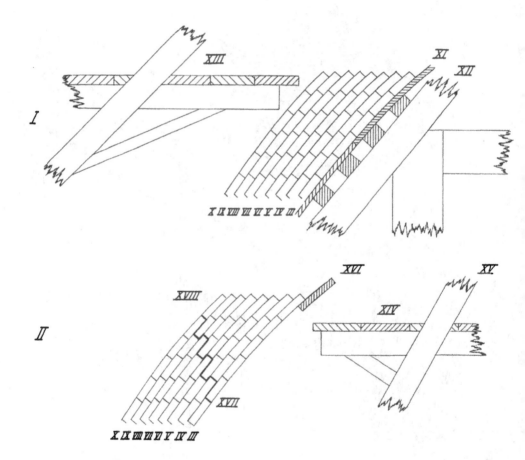

For his "designs and labors" he received 10 florins in 1417. Directly thereafter the Dome documents refer to presentation of models by artisans of Florence and Pisa. The impression arises that Filippo ordered model work from artisans and gave the idea of such work to others. Advice was also obtained from a man called Giovanni dell' Abaco. One of the models represented an armature and was made for Giovanni d'Ambrogio, the architect.[6]

This model work of Giovanni did not seem to delay him in the execution of his preestablished, Gothic-inspired plans. In the summer of 1418, he purchased logs for the Opera, "to make scaffolds for the vaulting of the Cupola."[7] Previously, when building the Tambour, he could use the inside scaffolds, already mentioned, but when now planning construction of the Cupola and no doubt relying on the use of a conventional centering, he needed new, outer scaffolds. It is not recorded that the wooden logs were actually used for this purpose. The debates about Filippo became all-absorbing.

On 19 August 1418, the Opera announced a major competition for further models. According to the biographers Filippo had suggested this step. The announcement was as follows:

... Whoever desires to make any model or design for the vaulting of the main Cupola of the Dome under construction by said Opera—for an armature, scaffold or other thing or any lifting device pertaining to the con-

struction and perfection of said Cupola or vault,—shall do so before the end of the month of September. ... If ... the model ... be used ... he shall be entitled to a payment of 200 gold Florins, and if any one does any work in connection with this matter the Opera will ... compensate him. ...[8]

This produced an avalanche of design proposals. In September and October of the year, work was done on a masonry model of Filippo, while Ghiberti submitted his own models. About a dozen competitors presented some seventeen additional models.[9] Attention was centered on Filippo. He had proposed to omit the centering armature, and he had novel ideas for the masonry design. He proved his ideas by his model, which the Opera built for him in order to permit "inspecting the work while in progress" and to determine "whether it is feasible." When the model was finished at the end of Octo-

[6] *Cupola*, Docs. 16, 21–28; *Manetti*, pp. 31 f.; *Vasari*, pp. 205 f.
[7] *Cupola*, Doc. 237.

[8] *Cupola*, Docs. 11, 21–42; *Manetti*, pp. 31 f.; *Vasari*, pp. 205 f.
[9] *Cupola*, Docs. 21–42. This set includes models by Ghiberti (Docs. 29 f.), Pesello (Doc. 42) and others. Several artisans presented both models of a cupola and of an armature for its centering. Leonarduzzo made a model of a hoist (Doc. 30). As to Filippo's models, the biographers' accounts are confused. *Manetti*, p. 44, speaks only of a small wood cupola model, *cupoletta piccola di legname*, supposedly made by or for Filippo in 1420. *Vasari*, p. 213, substantially repeats this view. He adds that it was made by Bartolommeo, a carpenter who lived near the Studio. According to Doc. 50, Bartolommeo di Marcho and a helper provided trim for the masonry model in 1419–1420 but did not build a wood model.

8. Brunelleschian scaffold for vaulting a hemispherical cupola without armature. Facsimile after *Nelli*, 1755, who adds a legend reading as follows in translation:

This demonstration is by Filippo Brunelleschi, architect, and was made for the scaffolds of the Cupola of S. Maria del Fiore, Florence, in the year 1419. It was what he showed when he was left free to be alone in the Cupola work, without Ghiberti his associate. He had not wished to reveal it before, since he was not free as architect to the Opera, as may be read in his Life written by various authors.

In fact the drawing shows a Roman-inspired cupola profile, similar to that of the Pazzi Chapel and S. Spirito. The legend is in error if it means that Brunelleschi kept the drawing secret until after 1420; by then the scaffold was under construction and everybody could see it. There is no independent corroboration that Brunelleschi made the drawing or a prototype for it, nor is it easy to think of others, prior to 1420, who would have made it.

The drawing uses fairly subtle perspective, combined with attempted orthogonal projection in cross-sectional view. Successive improvements of the latter technique may be found in Figures 31 and 35.

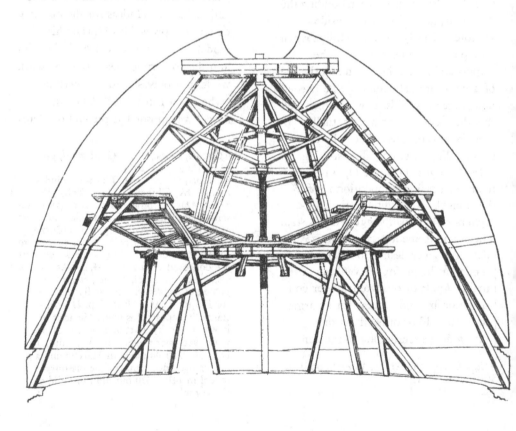

ber 1418, Giovanni d'Ambrogio was dismissed as chief architect.[10]

However, the time of unconditional approval for Filippo's plan had not yet arrived. In November the Opera appointed an acting architect, Antonio di Battista. It held two meetings in December, during which time the competitors were to "demonstrate and defend" their models.[11] These meetings, and the year ensuing, brought no decisive action beyond the harsh and negative step of dismissing the old architect.

Some modern writers say that Brunelleschi himself was undecided; that he made widely divergent models for the dome; and that he joined Ghiberti in making new models.[12] These writers are influenced by Guasti, the editor of the documents, who could not decide how many models Brunelleschi built, or how they differed from one another.[13] In our opinion there is no basis for the view that Brunelleschi showed signs of indecision or that Ghiberti aided him. The documents make it much more probable that each rival presented one or two preliminary models and a final model, all oriented toward one concept for each proponent, but not necessarily oriented toward the same concept for both, or toward converging concepts. Of course it is probable that the designs of the rivals matured, perhaps beginning as early as 1404 when they first

actively concerned themselves with Opera problems, but it does not appear that they were closer to one another in 1420 than ever during the previous twenty years. Various contestants submitted a succession of models for the Cupola and also for armatures and other woodwork.[14] So did Lorenzo, and it does not follow that he influenced Brunelleschi thereby, or that he was influenced himself.[15] An insistence on disagreeing with the biographers has led some writers to the opposite view, but although such insistence has been widespread since the discovery of the official documents, there is no good basis for it. The documents confirm the biographers much more often than they correct them.

The documents confirm, in particular, the biographers' story about the outspoken unpleasantness between Filippo and Lorenzo.[16] Clearly this had resulted from their rival claims for recognition as leading artists of Florence—a regrettable and unprofitable rivalry, but one that cannot be denied or explained away. Each rival of course had groups of vocal supporters among the Florentines, and the battle of these factions in turn caused both artists to engage in the intrigues and counterintrigues described in the biographies. The Opera showed that it wanted the two artists to cooperate, but there is no evidence that it was so successful in this case as it had been in the battle between Ghini and Neri.

[10] *Costruzione*, Doc. 478 f.
[11] *Cupola*, Docs. 15, 34.
[12] *Nardini*, p. 45; Krautheimer, *Ghiberti*, p. 90; P. Sanpaolesi, *La cupola di Santa Maria del Fiore. Il progetto . . .* , Rome, 1941, p. 8 (hereafter: *Il progetto*); G. C. Argan, *Brunelleschi*, Milan, 1955, pp. 56–63; *Sanpaolesi*, pp. 54 f.
[13] *Cupola*, pp. 192, 198.

[14] See, for example, *Cupola*, Docs. 39, 40.
[15] So also Frey, *La vite*, pp. 165 ff., who disagrees with Guasti.
[16] *Cupola*, Doc. 74 and pp. 195 f.

There is occasional reference in the documents to a "new model," but at no point does the context indicate that it was a model showing a change of earlier Brunelleschian plans or showing an influence of Lorenzo. On occasion, Lorenzo supplied some minor brass parts for Filippo's model, or later some larger castings for his machines, but during the period of planning and model building these contributions were minimal and formal.[17] Lorenzo made these contributions in 1419, when Filippo was about to persuade the people that the Cupola could and should be vaulted without armature. Filippo had proved his point about this procedure by constructing small cupolas elsewhere, using his method.[18] Lorenzo then condescended to supply some brass tacks for the project and to sign his name to the specification.

THE MASONRY MODEL

Near the end of 1418, four masons of the Opera had constructed Filippo's model, without armature, in about ninety days, not far from the Campanile.[19] In July 1419, Filippo was paid for a Lantern that he had placed on top of the model and for a Gallery between its Cupola and Tambour. He may have personally carved these parts of wood. Unfortunately they are lost and are hardly remembered in the literature.

It seems that he finished the model with care. Its Lantern had the Florentine emblem on a flag, *bandieretta col giglio*. Some part was lined with gilt tinfoil; perhaps it was applied to the inside of the model vault, where mosaics were proposed for the actual building. Early in 1420, Filippo provided wire and string to demonstrate measurements. One Stefano del Nero then made a painting of the model.[20] This painting is also lost. We can only estimate the size of the model from a few technical data. The masons used forty-nine horse loads of quicklime for making the cement needed for the model and another structure (a wall around the garden of an Opera official). A wood turner provided the circular windows set into the model Tambour and used for this purpose a wooden log, the size of which is reported. It appears that the model probably had a clear inside diameter, from wall to wall, of about 11 to 12 feet.[21] In other words, it

17 *Cupola*, Doc. 50: *Soldi 11 per aghutuzi/portò Lorenzo.*
18 *Manetti*, p. 37; *Vasari*, pp. 211 f.
19 *Cupola*, Docs. 18, 43. The basic instructions were given by Filippo, Donatello, and Giovanni d'Antonio di Bancho (Doc. 44). Among the supervisors (*maestri*) were Giovanni del Abaco, Pesello, Acquettini, Ricco di Giovanni, and Michele di Nicholò Schalchagna. The workmen were Andrea Berti Martignoni, Bonaiuto Pauli, Papio di Andrea, and Alioso.

20 *Cupola*, Docs. 20, 50.
21 About the quicklime see Frey, *Le vite*, p. 164; the log: *Cupola*, Doc. 50. The reported amount of lime provided cement for ca. 150 English tons of brick masonry. The size of the garden wall is unknown. Estimating that the model weighed 100 to 120 tons and that it stood on masonry piers ca. 7 feet high to provide access, we find that the Tambour probably had some 11 to 12 feet clear space from wall to wall. Master Chiaro used a "9½ foot tree trunk to make eight 'eyes' on the lathe." If he sawed this trunk into eight blocks ca. 1 foot thick, he could produce windows for a Tambour of the indicated size.

showed the proposed Cupola and its support on a scale of about 1:12. In its outer dimensions the model must have been about half as wide as the adjacent Campanile. The "small model" of 1367 was still standing near it, inside the church.

While building this Cupola model, Filippo began to revolutionize architecture by his *Innocenti* Arcade design. He did not introduce technical innovations into that design, or administrative changes into the Opera's methods. When his Cupola model had received its finishing touches, the Opera appointed Four Citizen Cupola Provisors, *cives pro constructione cupule;* these men gave final review to the models of all contenders, in April 1420, and then elected new superintendents for the Cupola. In this position they elected Filippo and, to his rage, Lorenzo. As an afterthought they also elected the acting architect, Antonio di Battista, as one of the superintendents.[22]

Soon after these appointments the Opera received a written description of Filippo's masonry model. This description as well as the masonry structure that it described was called *modellum*.[23] The text is generally assumed to be mainly, if not exclusively, of Filippo's composition, although it was also signed by Lorenzo and Antonio, and although no draft or copy in Filippo's handwriting has been found. It is a document of remarkable clarity and extreme conciseness; some of its technical terms are in need of explanation. The document exists in form of a notarial record preserved by the Woolen Guild, which generally states what is shown by the masonry model.[24] Other copies exist, apparently based on a lost paraphrase, given by the Guild to its Opera, which generally says what "should be" done in the actual construction.[25] In substance the two versions are identical.

It is clear from the text that the document resulted from careful elaboration of inventions and deliberations which no one could formulate in such conciseness without weeks or even years of work. A different but implausible picture is drawn by Vasari, who describes the composition of the document as a spontaneous act and rapid performance of Filippo; the contents given by Vasari are garbled. It is one of his weaknesses that he tries to invent dramatic action, in imitation of Thucydides but with inferior means. The pertinent account of Manetti is much more useful at this point, as it makes no attempt to introduce speeches or actions.

The Opera has not preserved its copy of the precious document, but Manetti

[22] *Cupola*, Docs. 1, 2, 71; *Brunelleschiana*, pp. 14 f.
[23] *Modello facto per esempro della cupola; modelli . . . scripti.* Also see Acquettini's "parchment model," *Cupola*, Doc. 181.

[24] Arte di Lana. Partiti, atti e sentenze, vol. 149, fols. 59v, 60r, Archivio di Stato, Florence. The document was discovered by A. Doren: "Zum Bau der Florentiner Domkuppel," *Repertorium für Kunstwissenschaft*, XXI, 1898, pp. 248 ff. Also see *Brunelleschiana*, pp. 16–19.
[25] *Copia del capitolo tratto duno libro dell opera di Sta. Maria del Fiore*, in Magliabecchiano XIII 72, fols. 37v, 38r, Biblioteca Nazionale, Florence; *Manetti*, pp. 38–41; *Vasari*, pp. 209–211; the latter two are copied in *Cupola*, Doc. 51.

saw it and copied it. The Opera documents still contain specifications that are amendments made in later years.[26]

THE STRUCTURAL PLAN

Filippo did not find it necessary in this specification to mention the diameter of his Cupola. This dimension was fixed by the diameter of the octagon, then existing, which in turn was based on the decision of 1367. Proceeding on this basis, Filippo set down the form and measure of all component parts and the functions of the major parts. Later, when building the Cupola, he complied with the specification in almost every detail, as is most clearly noted by a study of measurements, made by Nelli about halfway between Filippo's time and ours.[27] The basic elements are indicated in our Figure 9. Filippo introduced them with a short preface:

Hereafter we will particularly record all parts contained in this model, constructed as a likeness of the large Cupola.

He then gave a three-point description of the Cupola shells:[28]

[1] First, the inner cupola is vaulted in five-part form in the corners.[29] Its thickness at the bottom point from which it springs is 7 feet.[30] It tapers so that the end portion surrounding the upper oculus is only 5 feet thick.

[2] A second, outer cupola is placed over this one to preserve it from the weather and to vault it in more magnificent and swelling form. It is 2½ feet thick at the bottom point from which it springs, and tapers to the upper oculus, where it is only 1¼ foot thick.

[3] The empty space between the two cupolas measures 4 feet at the bottom. This space contains the stairs to give access to all parts between the two cupolas. This space terminates at the upper oculus, 4⅔ feet wide.

Nothing was said about the Lantern standing on this oculus in the model. Filippo may have felt, correctly, that this element is structurally unimportant so long as its weight is not excessive and that its stylistic design required further elaboration.[31]

---

[26] *Cupola*, Docs. 52, 75. For the three specifications see our Document section, *infra*, pp. 139 ff.

[27] G. B. Nelli, *Piante ed alzati interiori ed esteriori dell' insigne chiesa di S. Maria del Fiore metropolitana fiorentina*, Florence 1755, ed. G. B. C. Nelli (hereafter: *Nelli*). The drawings were also published by Nelli's draftsman: B. S. Sgrilli, *Descrizione e studi dell'insigne fabbrica di S. Maria del Fiore*, Florence, 1733. Later they were copied by G. Molini and G. Del Rosso, *La metropolitana fiorentina*, Florence, 1820, and—in part—by W. B. Parsons, *Engineers and Engineering in the Renaissance*, Baltimore, 1939, pp. 602 f., 605 f.

[28] The numbers in square brackets are used by *Cupola, Brunelleschiana*, and other modern publications, not the manuscripts. Basic points for a structural analysis: Parsons, *Engineers*, pp. 594–599.

[29] This is the usual interpretation; see for example *Vasari*, p. 209, "the inner part." However, it may be more accurate to translate: "The Cupola is vaulted in five-part form in the inner corners." About the actual curvature see P. Sanpaolesi, "Il rilievo della cupola del Duomo di Firenze," *Rivista d'arte*, 1937, p. 80.

[30] This is the thickness measured along a diagonal of the octagon, as are the dimensions specified in points 2 and 3.

[31] It is even conceivable that no definite

The five points of the specification, following in next place, relate to the reinforcements of the Cupola and the materials used:

[4] There are 24 ribs (*sproni*), 8 in the corners and 16 in the sides. Each corner rib has a thickness of 14 feet at the outside. Between the corners there are 2 ribs in each side, each 8 feet thick at the bottom.[32] The ribs tie the 2 vaults together. They converge proportionally to the top, where the oculus is.

[5] The said twenty-four ribs, with the said cupolas, are girdled by six circles (*cerchi*) of strong sandstone blocks.[33] These blocks are long, and are well linked by tin-plated iron.[34] Above said blocks are chain rods of iron (*catene di ferro*), all around said vaults and their ribs. At the start solid masonry has to be laid, 10½ feet high,[35] then the rib [outlines] must be followed [separately].[36]

[6] The first and second circles are 4 feet high, the third and fourth circles are 2⅔ feet high, the fifth and sixth circles 2 feet high, but the first circle, on the bottom, is also reinforced with long sandstone blocks laid transversely, so that the inner and outer cupolas rest on said blocks.[37]

[7] At the height of every 23 feet or thereabout, of said vaults, there will be small barrel vaults (*volticciuole a botti*) from one [corner] rib to the next [rib], going around said cupolas. Below said [system of] small vaults from one rib to the other are big oak beams (*catene di quercia*), which tie the said ribs. Above [each of] said timbers is a chain rod of iron (*catena di ferro*).

[8] The ribs are entirely built of grey and tan sandstone,[38] and the covers or

---

decision has been made to provide the Cupola, like the model, with a lantern. We owe this recognition to A. Parronchi. A popular tradition, recorded by *Vasari*, p. 222, has it that the weight of a lantern was needed for some structural reason. There is no valid basis for such an opinion. Also note that in 1601 large parts of the Lantern were removed and then replaced by Buontalenti and Bronzino, and the Cupola did not suffer. *Cupola*, Docs. 363–381.

[32] This dimension was modified in March 1420–1421 "to remove useless weight," *Cupola*, Doc. 52. Evidently Brunelleschi did not share the popular superstition recorded by Vasari. Parsons, *Engineers*, p. 595, is somewhat critical about Brunelleschi's use of intermediate ribs. The point is too complex for discussion in this study.

[33] *Macigni*, blocks of a grey or light-blue sandstone quarried in hills north of Florence. Its use is a characteristic of Brunelleschian architecture. F. Rodolico, *Le pietre delle città d'Italia*, Florence, 1953, pp. 237 ff. The "six circles" were changed to seven, in March 1420–1421; one circle was added in the bottom region.

[34] Actually, lead-lined iron was used: *Il progetto*, pp. 16 ff. The iron chain rods lie "around" the Cupola like the hour marks of a clock dial, not in peripheral alignment.

[35] Actually, solid masonry extends up to ca. 20 feet above the upper, inner walkway around the Tambour.

[36] *Assi . . . poi seguire i sproni*. In *Osiris*, IX, 479, 485, Prager interpreted this that the ribs were separate and largely of stone, the shells of brick. According to a personal communication of Dr. Saalman the ribs are homogeneous with the shells.

[37] The outer ends of these blocks, and somewhat higher ones added in 1421–1422, are visible; see Figure 10 at "1." Equally visible are the parts specified in the next paragraph.

[38] *Pietra forte*, a tan sandstone quarried in hills south of Florence. Rodolico, *Le pietre*, pp. 239 f.

faces of the cupolas are entirely of tan sandstone, tied to the ribs, up to the height of 46 feet.[39] From there on upwards the masonry will consist of brick or porous stone, as may be decided by the man who has to build it then. At any rate it will be of material lighter than rock.[40]

It will be seen that Filippo relied on the use of upright ribs interconnected by several types of horizontal members: peripheral tie rings of stone, wood, and again stone, and "small" or arcuate vaults in the corner areas above the wooden tie ring. With regard to some details such as the bricks, the plan was tentative, but it was set forth in definite terms, as provided by Gothic practice. The upright members are consistently called *sproni*, and with equal consistency the peripheral stone tie circles are called *cerchi*. For connecting elements the specification always uses the word *catene*.[41] These elements were to

[39] In March 1421–1422, the specified interface between stone and brick was lowered to a level 22 feet above the Tambour, as another measure "to remove excessive weight."

[40] Amplified in 1425–1426: "Large Bricks shall be made . . . They shall be placed in such herringbone pattern as is deliberated by the master in charge." The pattern is shown in *Sanpaolesi*, Pls. 28, 30.

[41] However, there are different types of such elements, and the word also means "chain" in the sense of a series of links. In the second sentence of point 7, *catene* occurs in three senses: there is a *catena* or linked series of forty-eight *catene* or *legni*, timbers, held to the masonry by *catene ferri*, iron rods. The latter term occurs also in paragraph 5. The several meanings of *catena* are mentioned for example in

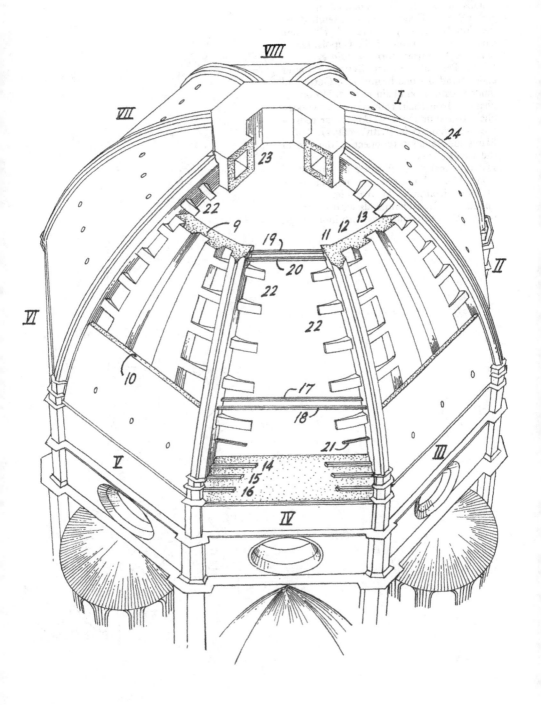

35

10. Brunelleschi's Chain System and its visible parts. Drawn by G. Rich, 1969. *I.* Stone Chains at one of the higher levels. Details: see *IV. II.* The Wood Chain (after *Nelli*). Details: see *V* (sketched in the Cupola walkway by Prager). *III.* Stone Chains in the bottom of the Cupola. Details: see *VI* (after *Forma e colore*).

Visible Parts of the Chain System. *1.* Outer ends of radial stone beams. *2, 3.* Outer and inner ends of iron chain rods. *4.* Wood chain. *5.* Iron chain rods tying the wooden chain to the masonry. *6.* Connector rods of the iron chain rods (with wedges). *7, 8.* Plates and bolts interconnecting the logs of the wood chain.

Hypothetical Parts. *9.* Inner tie ring of stone. *10–13.* Additional tie rings of stone. *14.* Their iron links. *15.* Iron chain rods (plain). *16.* Iron chain rods with outer hooks. *17.* Iron chain rods with inner rings.

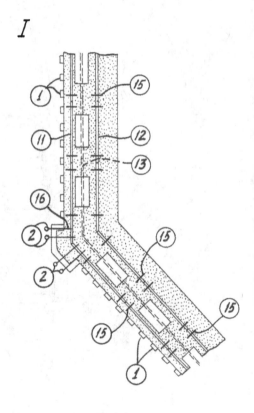

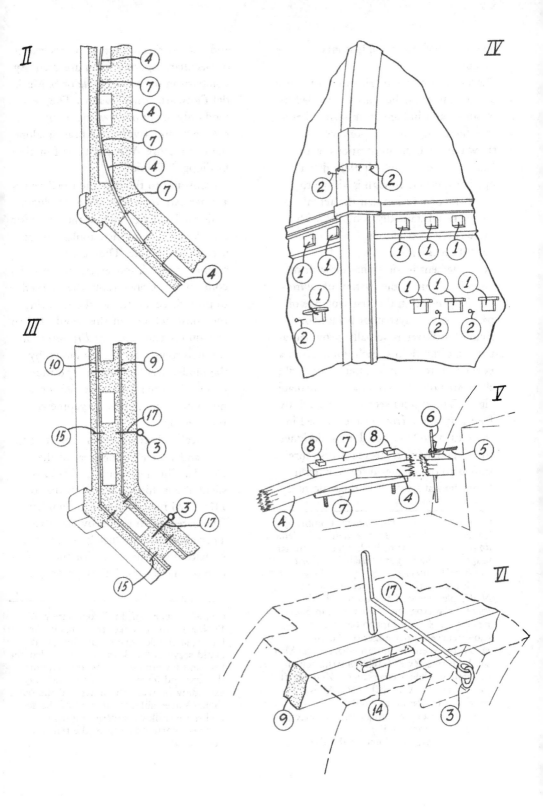

37

hold the tie rings to the masonry (Figure 10).

Large parts of the tie rings and of their connectors or anchors are embedded in masonry and hidden from sight; therefore these rings and anchors were not shown in the basic illustrations of the Cupola structure, which Nelli developed. However, enough is visible to allow the conclusion that the actual structure substantially complies with the written instructions. It was a successful specification.

One element is somewhat difficult to visualize. This is the system of *voltic-ciuole* specified in the first sentence of paragraph 7. A system of bridges or flying buttresses is actually provided at the indicated points, each going from a corner rib to the next intermediate rib; there are none between the intermediate ribs.[42] The system seems developed, by free adaptation, from elements used in the Battistero.[43] Other domes, designed by Filippo, have reinforcements more similar to the medieval form, known from the Battistero.[44] In those domes,

and the Battistero, actual barrel vaults of peculiar, tapering form are used. By comparison the *volticciuole* of S. Maria del Fiore are "small" vaults. They extend only about one *braccio* along their axis and a very few braccia along their curvature. What is their function in Filippo's Dome?

According to the text the small vaults are *per andito intorno alle decte cupole*. This can hardly mean that they are "for a walkway around the cupolas," to be used by the public. They are too steeply inclined, too short, and too disconnected.[45] Conceivably they served to provide access for workers, during the construction, but this would fail to explain the expression *andito intorno*. It seems to us that they, as shown by the model, were "going around the cupola" and that the expression does not state a function but a geometric relationship.

Statically these elements serve to support and stiffen corner areas of the Cupola. The specification of 1420 is silent about this function, but that of 1425–1426 may be found to confirm it. This later document provided that, at a certain elevation, "brick masonry shall be laid in arc-shape . . . for the perfection of the tie ring." This masonry is

---

Cambridge Italian Dictionary, Cambridge, England, 1962, and *Grande dizionario della lingua italiana*, Turin, 1962. There is similar usage as to the English "chain." *Oxford English Dictionary*.

[42] This is clearly and correctly shown in *Nelli's* plan view of the Cupola. In this respect the axonometric diagram in *Sanpaolesi*, Pl. 42, appears to be in error, as it shows curved inner edges of the *volticciuole* between the intermediate ribs. Also see J. Durm, "Zwei Grosskonstruktionen der italienischen Renaissance," *Zeitschrift für Bauwesen*, XXXVII, 1887, Atlas, reprinted by Parsons, p. 604.

[43] See for example Sanpaolesi and Bucci, *Duomo e Battistero*, Fig. 4.

[44] For example, the dome of the Pazzi

Chapel: *Sanpaolesi*, Pl. E, opposite p. 65.

[45] Nor are there other structures in the Dome, at the designated locations, that could serve as "walkways," except that the first and lowermost *volticciuole* support the first and lower-most of the walkways extending between the inner and the outer shell. Conceivably, ten instead of the actual three walkways were originally planned, corresponding to the ten levels of *volticciuole*.

likely to be one set of *volticciuole*, at the level that was then being constructed.[46]

Sections 9 to 11 differ from the rest of the specification in that they relate primarily to external features, not the structural system of the Cupola. At least one of these features, the curved corner ribs of the Cupola, is visible far and wide and is shown in our Figure 11. Another feature, a gallery between the Cupola and the Tambour, was not actually built, and still another, providing drain spouts, would be needed only in the presence of the gallery. The masonry model showed the gallery, but we do not know how it was constructed. We only read in the specification that it was to harmonize with "the little tribunes farther down," which Filippo actually built in the late 1430s and early 1440s.

[9] There shall be an outer walk above the lower, eight round windows,[47] supported on brackets, with parapets, perforated and about 4 feet high.[48] [It shall be made] in harmonious relationship with the little tribunes farther down, or really [there shall be] two walks, one above the other, above a well-ornamented cornice. The upper walk shall be open.

[10] The water falling on the Cupola collects in a marble gutter two thirds of a foot wide, which discharges the water by suitable drain-spouts of tan sandstone below the gutter.

[11] There shall be 8 ridges of marble,[49] on the corners in the surface of the outer cupola. They shall be as wide as indicated [by the model] and shall rise 2 feet above the Cupola. They shall be contoured,[50] in roof-like design, 4 feet wide on top, so that there is 2 feet from the peak to the side edge everywhere. They shall converge from the springing line of the vault to the [upper] end.

Another explicit remark of Filippo's specification, dealing with aesthetic appearance, may of course be found in the provision of an outer cupola in "magnificent and swelling" form, paragraph 2. That remark, as well as the section dealing with the marble ridges, is generally known, but the literature has strangely overlooked or disregarded the section dealing with the

[46] The Cupola confirms this, as it uses a peculiar sequence of elements "22." They are disposed at vertical distances of ca. 12, 11, 11, 11, 11, 13, 9, 9, 9, and 9 *braccia*. *Nelli*, Figure IX. (Our own Figure 9 is diagrammatic and shows only 7 levels.) The odd spacing of 13 *braccia* occurs exactly at the indicated level.
[47] *Sopra gli otto occhi di sotto*. The expression indicates that the model also had the "upper" and smaller windows, which illuminate and ventilate the walkways. They are perspectively in line with the several tie rings of wood and stone, whereby they give expression to these upper, structural ties.

[48] *Braccia 2 o circa* (Arte di Lana record) or *braccia 2 in circa* (*Manetti* and *Copia del capitolo*). It is conceivable that Filippo's draft said *braccia 20 circa*.
[49] The marble came from Carrara. Rodolico, *Le pietre*, pp. 241 f.
[50] *Scorniciate*. In modern Italian this means "without border or edge." In Quattrocento Italian it meant rather "with border or edge."

11. S. Maria del Fiore. Photo Alinari

outer walkways and the Exedras.[51] Some elements of the section may be hard to interpret, but so much is certain that Filippo, in his model and specification of 1420, proposed an external Gallery and Exedra structure of unitary and harmonious design. As indicated in our notes to section 9, it seems possible that Filippo wanted the Gallery part of this structure to be as high as the Tambour. In any event he wanted the entire structure above the lower walk of Talenti to be in his own style, not the actual pseudo-Trecento style applied by his successors. Michelangelo sensed Filippo's plan, but was only able to stop the construction of Baccio's gallery by ridiculing it as a cricket cage.

The specification ends with a brief section wherein the method of vaulting without armature is proposed for the lower one-third of the Cupola. Up to that height the use or nonuse of a centering, spanning the vault, was not critical, since this lower part of the shell is nearly vertical and the successive rings of masonry are self-supporting by simply resting on the lower rings.

However, section 12 indicates that internal platforms be constructed. These were known to the workmen from the construction of Filippo's model. They occupied the space needed for the armature of a centering, if a centering were used. Accordingly, the provision can be paraphrased by saying that it revocably committed the Opera to the use of Filippo's method. Apparently the scaffolds planned by Giovanni d'Ambrogio in 1418 had not been built. [12] The Cupola shall be built in the aforesaid manner, without any armature, at least up to a height of 58 feet, but with platforms in such manner as will be counselled and deliberated by the masters who will have to construct it, and from 58 feet upwards, as will then be counselled, because in masonry work practice will teach how to carry it out.

We have no account, other than Vasari's fictional story, of what was said in the Guild Hall when this specification was read by or for Filippo. Nor do we know much about later Quattrocento references to this specification, with the exception of certain amendments, described in the course of our technical explanation. However, there is evidence that qualified builders recognized the importance of the architectural system described in the document. A statement, later garbled by scribes but still recognizable as a brief version of Filippo's specification, occurs in one of the technical instruction books composed in the later part of the century.[52] However, the more philosoph-

---

[51] *Brunelleschiana*, p. 20; Heydenreich, "Spätwerke Brunelleschis," p. 14; and *Sanpaolesi*, pp. 53 ff. R. Sabatini, *Uno studio per la completazione del tamburo della cupola*, Florence, 1943, interprets point 9 as calling for a *2-braccia* parapet; this may be accepted in our opinion. He then continues his interpretation by saying that the parapet was to be similar to that of Talenti, below the Exedras. However, the specification speaks of similarity between Exedras and Gallery, not between parapets. Some suggestions about the present point may also be found in *Nardini*, p. 76. In our opinion they are more to the point than the current views about Filippo's "Spätwerke."

[52] *Trattati*, pp. 92, 93.

ical architecture books, beginning with that of Alberti, disregard it.

BUILDING THE CUPOLA

At the end of July 1420, the authorities accepted Filippo's model and specification. The document was also signed by Lorenzo, and of course by Antonio. Filippo then at once proceeded to build the Cupola. He supervised the work done by carpenters, stonemasons, and bricklayers, sent messengers to stone quarries, contracted both personally and through others for the supply of materials, and built a powerful hoisting machine. The documents are not explicit about the authorship of various elevated cranes and stone-positioning devices that cooperated with the ground-supported hoist. Only secondary tradition connects Filippo's name with such details as the design of certain hangers and hooks, whereby the devices manipulated the heavy loads. With regard to all this machinery, Antonio stayed entirely in the background, and Lorenzo appears less often than some minor contemporary workers. The documents speak constantly of Filippo, sometimes of individual ox drivers or makers of specified parts, and almost never of the two co-appointees or their substitutes, with regard to the system of machines and the individual devices. We will analyze the documents, as well as pertinent drawings, in a separate study because they provide a large, self-contained record that deserves separate consideration and may, when clearly understood, aid us in our review of other documents on Quattrocento building activities.

In 1421, Filippo received a patent for his transport ship, as well as a large prize for his hoisting machine. In 1423, he received another large prize, once more for the machine and also for the chain structure used in the Cupola itself. The two prizes amounted to 200 florins, the sum originally promised to the winner of the Cupola competition. Evidently the Opera tried to designate Filippo as the winner of the competition, although it, at the time of these awards and for another year or two, also retained Lorenzo as building master with equal rank or powers, thereby irritating Filippo.

Not every part of Filippo's work was a clear success. There is uncertainty and confusion, at least in his biographers' accounts, about his wooden chain. The transport ship was a failure, and the biographers tried to suppress its story, as has been shown in another study.

A few years after the start of the Cupola construction, when the work approached the 58-foot level that called for further considerations (Point 12 of the specification), the friction between Filippo and Lorenzo became very strong.[53] A faction that identified itself with Lorenzo argued for providing more light, and proposed that twenty-four large spaces in the lower part of the cupola, between the ribs, should be

[53] *Manetti*, pp. 46 ff. An outbreak of the plague, in 1424, may have caused additional irritation.

left open to provide windows. Filippo rejected the proposal. He insisted on a safe, solid structure, such as he understood. He adhered to this principle although it made the vault rather dark, as it is illuminated only through the limited area of the Tambour windows that he or others had provided in the previous decade. The debate reached its height in the winter of 1425–1426, when preparations were made for the second layer of tie rings and their supports and wall clamps or chain rods. This was the last moment for decisions about omission or removal of masonry between the ribs, below these rings. The Opera then dismissed Lorenzo for several months. Lorenzo was reappointed later, but with sharply reduced responsibilities and corresponding powers.[54]

We will explain the technical problems of these tie rings and chain rods in a separate chapter. On a human plane, the biographers make Lorenzo's dismissal very plausible, although they do not give detailed references to the documents. They resolve the affair into a story of intrigues and counter-intrigues. Although they favor Filippo, and even despise Lorenzo, they do not mince any words about Filippo's own devious practices. Every historically and technically significant part of the story is supported by the official record. It is obvious that Filippo often used expressions of violent hatred. Lorenzo's resentment appears in a late, self-serving misstatement of fact.[55] The one man

felt unappreciated as a sculptor, the other as an architect.

Nevertheless the Guild was able to secure consensus when it was needed or to have this achieved by the Opera or its Four Citizens as Cupola Supervisors. For example, a major addition to the original specification was adopted in 1426, when technical details of the program had become problematic. Among these details was mainly the question about omission of windows piercing the Cupola walls. The original plan relating to this point was confirmed by the new specification. More specific provisions were added. Lorenzo signed the document without further debate, so far as the record shows. Filippo agreed, perhaps as a concession to the window faction, that in addition to the series of openings in the outer shell (which vent and illuminate the inner walkway near the level of each pair of tie rings) there be provided a set of inner, round openings. These openings would in due course provide access for mosaic workers. However, for the moment these inner openings were to be filled with masonry, avoiding local stresses with maximum assurance. The new specification also provided details for elements of reinforcing brickwork, perhaps in reply to

[54] *Cupola*, Doc. 74 and pp. 195 f.
[55] With regard to Brunelleschi, the Opera decided in 1436 (*Cupola*, Doc. 273) "to

have him come to the office and to ask him, in such terms as are . . . required for him kindly to defer all rancors that he still harbors." With regard to Brunelleschi's "difficult associate" (*Sanpaolesi*, p. 11), we refer to the self-serving and inaccurate statement at the end of the second of the *Commentaries*: "In building the Cupola, Filippo and I cooperated for 18 years, at one and the same salary."

questions raised by artisans. Other parts of the fabric, by contrast, were re-designed to make upper portions of the Cupola lighter and thereby further to protect the shells from bursting. The new specification also fixed specific dimensions for bricks, and mentioned bricklaying patterns, as well as details of the scaffolding and tooling. It continued the use of vaulting without armature, which at this point began to be important.[56]

In the six years that followed (1426–1432), ring after ring of wall and rib masonry was placed, closed, and made self-supporting. Filippo had this done by the work of 40 to 50 masters,[57] with the help of contractors and helpers who brought the materials into the work yard, guided the oxen to operate the hoist, and performed other secondary tasks. When modern builders construct a cupola of comparable size, from similar materials, they generally use lighter construction and are ready to pierce the cupola by large windows, which is possible in the light of experience that Filippo and others provided in later, smaller cupolas.

In order to make the structure strong and durable, the modern builder uses principles similar to those of Filippo's model, specification, and Cupola. He generally uses hollow cells, similar in substance to the cells provided by Filippo (Parts 1 to 4 of his specification). Each cell will have inner and outer walls (Parts 1 and 2) and upright sides (Part 4). It will have horizontal reinforcements (Parts 5 to 7). If the builder is architecturally candid, he will give visible expression to each major part, as Filippo proposed to do (Parts 9, 11). He may or may not use vaulting without armature (Part 12).

LATER DESIGNS

People who had seen the hesitations of earlier Dome architects were deeply impressed by Filippo's self-assurance. They responded with a surge of enthusiasm, such as few architects ever provoked.[58]

The Opera made further attempts to achieve cooperation between its famous architect and the equally famous gold-smith and bronze founder, Ghiberti, who worked for the administration of the Battistero and other clients. A model of the entire cathedral, ap-proved by both men, was produced in 1429 and shown in 1430 to the guild consuls and others. Unfortunately the documents contain only formal state-ments about it.[59] They create the impression that it was soon forgotten.

Meanwhile Filippo gave final review to the visible parts of the cathedral structure that his model had shown in outline and his specification had

---

[56] Also see *Vasari*, p. 220, and *Sanpaolesi*, pp. 55 ff., 106 ff., who report popular stories about "eating houses" maintained on the Cupola scaffolds. About the vaulting method see our Document section, *infra*, p. 142.

[57] *Il progetto*, p. 21. About labor forces employed in other Gothic constructions, see M. S. Briggs, *The Architect in History*, Oxford, 1927, p. 85.

[58] *Manetti*, p. 65; *Fabriczy*, pp. 151 ff., 354 ff.; *Sanpaolesi*, pp. 110 ff.
[59] *Cupola*, Docs. 61–70.

described in general. He designed and built his four Exedras, creations of the purest Renaissance style, between the larger half domes of the Trecento Tribunes. These Exedras were to be part of a larger, externally harmonizing system, intended to cover the entire, overlying structure. Next to the Exedras, at the corners of the octagon, large ridges project from the masonry, rising to the Tambour and Cupola and all the way to the Lantern. There the ridges are united by a structure that has strong Gothic accents. Whether the artists of 1367 would have liked it or not, Filippo used posts and scrolls on the Lantern, reminiscent of flying buttresses. He also used here the design of classic post and pilaster structures.[60]

Inside his Cupola, Filippo fully complied with the injunction of 1367, as he totally avoided any use of tie rods or tie beams spanning his vault or otherwise "visible" beyond its surfaces. He did here what the artists had hoped for and what Giovanni d'Ambrogio had been unable to visualize without a complete return to the Gothic past. However, Filippo knew better than to avoid all "visible" showing of the structure that he used. The outer, curved white ridges running over the Cupola, and downwardly continued to the bottom level of the Exedras, force-

fully mark the rigid corner rib structures. The light and vent apertures in the tile work surfaces of the Cupola indicate, perspectively seen from the ground, the levels of the wooden chain and the middle and upper tie rings. We think he also intended more forceful marking of the lowermost and major tie ring, but did not live to finish this program.

Some of his works were finished by others. The Florentines revered his memory and were disturbed when epigones introduced innovations and errors, as they finished his work on the churches of San Lorenzo and Santo Spirito and other buildings that he had begun. He had taught his perspective method to painters, sculptors, and intarsia workers, who in turn taught it to mechanical draftsmen.[61] However, he could not provide his followers with his genius and his sensibility, and some of his works were spoiled, as a result.

Among the works that were spoiled is a most important part of the Tambour-Cupola structure, the part that could have given visible expression to his principal tie-ring system. As a visible part of this system, point 9 of the specification proposes a pair of *anditi*, external gallery elements, between the Tambour and the Cupola. They, with interconnecting gallery structure between them, would indicate the position and direction of the major tie chains, at the bottom of the Cupola. Perhaps they would accentuate the connections with corner ribs by the use of pinnacles or statues at the corners. The total design would harmonize with the

[60] The top or *palla* was finished by Verrocchio, with the help of young Leonardo, 1467–1472 (*Cupola*, Docs. 325 ff., pp. 203 f.). The characteristic scrolls in marble reappear in later works of Brunelleschi, Donatello, Alberti, Leonardo, Michelangelo, and others; see Frankl, *The Gothic*, pp. 264–270; R. Wittkower, *La cupola di San Pietro di Michelangelo*, Florence, 1964, Ch. V.

[61] *Brunelleschiana*, p. 3.

Exedras, placed farther down. However, the execution of the plan is
unfinished. In the time of Filippo's
successors, the architectural effect of
the Exedras was blurred by a super-
abundance of polychrome marble, all
around these structures. The large,
outer gallery was not even started.[62] In
its place, Baccio d'Agnolo built a
ridiculous "cricket cage," which
Michelangelo halted.[63] In structural
history, the tie rings were an epoch-
making success,[64] but the corresponding
"visible" element was forgotten.

[62] *Forma e colore*, Pls. 12, 25. Regarding
this time, also see *Manetti*, pp. 68–71, but
note that his reference to the Gallery is
garbled. *Vasari*, p. 221, clarified the matter
to some extent by a brief sentence.
[63] *Cupola*, Docs. 341 ff. There have been
various proposals to remove the "cricket
cage." In Molini-Rossi, *La metropolitana
fiorentina*, the authors propose the installation of a smooth iron band, ca. 3 feet
high (see their last plate, unnumbered).
Guasti opposed it, *Cupola*, p. 206.
[64] When Galileo had founded the science
of statics by his *Discorsi e dimostrazioni*,
Leyden, 1638, the structures of the major
cupolas were measured and described by
C. Fontana, *Templum vaticanum*, Rome,
1694, and by *Nelli*, ca. 1695–1725. The
static characteristics of the cupolas were
considered by these writers and, later, by
T. Le Seur, F. Jacquier, and R. J. Bosco-
vich, *Parere di tre matematici sopra i danni
che si sono trovati nella cupola di San
Pietro*, Rome, 1743; G. Poleni, *Memorie
istoriche della gran cupola del tempio
vaticano*, Padua, 1748. These works became the major beginnings of civil engineering, as is shown (although without
reference to Nelli) by Strauch, *History of
Civil Engineering*, pp. 62–117.

THE NEW CONCEPTS
FOR CONSTRUCTION

3 VAULTING WITHOUT ARMATURE
The most famous or at least most
sensational invention of Filippo,
vaulting without armature, is described
in rather different terms by Vasari and
others. According to some it was to
save expense; according to Vasari it was
to avoid a technique that could not
succeed. Which interpretation is right?

It was common or at least widespread
practice to employ centerings and
"armatures" for them. It was usual to
place vault masonry on a "centering,"[1]
that is, on a preformed, masonry-
supporting, armature-reinforced board
structure that remained in place until
the binding and at least most of the
shrinkage of the mortar was completed,
and then was carefully removed, leaving the masonry in self-supporting
condition. As centerings usually consist
of numerous boards, they need the
"armature,"[2] interconnecting these
boards and combining them into a rigid
structure: a truss that supports the
centering that supports the new
masonry. The centerings and their
armatures are either ground-supported
constructions, wherein the armature
part includes upstanding posts, planted
in the earth, or they are elevated and
suspended from preexisting masonry, in
which case they are also called hung or
flying centerings. In Brunelleschi's time,
both centering boards and armature

[1] Centina, centricum, cintre, Vorbogen,
lagging.
[2] Armadura, armatura, charpente, Bo-
gengestelle, Lehrgerüst.

beams were made of wood, and
usually there was straw and clay between the boards and the masonry. All
this was generally practiced everywhere.

Ancient and Gothic builders had also
used a more daring procedure, wherein
they first built free-standing ribs of
masonry, on centering armatures large
and strong enough only for this purpose. They then used these ribs—
generally a pair or larger number of
arches, intersecting centrally—as a
stone armature for flying centerings
supporting the wall masonry that they
placed between these ribs. About a
hundred years ago Viollet-le-Duc
rediscovered the fact that early builders
had used this remarkable process. The
discovery led him to lay some of the
first foundations to the modern method
of steel skeleton construction.[3] The
stone armatures could be called
"permanent armatures," as they remained in the masonry; however, since
the terminology of artisans is flexible,
the words armature and centering are
often equated, and the stone or permanent armatures are widely known as
"permanent centerings," *cintres perma-
nents.* They may be considered one of
the major inventions of architecture. In

[3] E. Viollet-le-Duc, *Dictionnaire raisonné
de l'architecture française du XIe au XVIe
siècle*, I–IV, Paris, 1854–1868, s.v. *armature,
chainage, charpente, construction;* A.
Choisy, *Histoire de l'architecture*, II,
Paris, 1899, *passim;* and J. Fitchen, *The
Construction of Gothic Cathedrals*, Ox-
ford, 1961, pp. 139 f., 191 and *passim.* For
popular and literary views, see Frankl,
The Gothic, pp. 145 ff. and *passim.* About
Viollet-le-Duc and skeleton construction:
S. Giedion, *Space, Time and Architecture*,
Cambridge, Mass., 1941, p. 140.

early times they liberated the builder of a large cupola or arch from the strict limitations, imposed by the dimensions of available trees, and the supplies of boarding. As techniques of making trusses were developed only much later, in the times of Palladio and his followers, long after Filippo, the use of a permanent stone armature was an almost absolute necessity in the building of vaults spanning for example 70 or 90 feet or, as the Florentine vault, spanning 140 feet. These problems no longer exist today, as steel armatures have become available, which can be built in almost any desired size; they are superior to stone because they are much lighter, and they are also superior to wood because they avoid hygroscopic shrinkage and other defects.

At this point we refer to Figure 7, in order to show at 7-II what "can" be done without centering armature, instead of the more conventional procedure 7-I. Nowadays the technique schematically illustrated at 7-II is sometimes called "freehand vaulting." It uses only an inside scaffold that is strong enough and is placed to support workers and the materials they employ during the day. Method 7-I uses a much stronger inside armature supporting the entire vault. Method 7-I therefore needs an outside scaffold for the workers and the materials brought up for placement. In other words, 7-II dispenses with the armature and places the scaffold elsewhere than in 7-I. The central placement of the scaffold greatly facilitates the work.

The freehand method does not necessarily dispense with centerings. On the contrary, the workmen proceeding according to this method often use small portable centering boards, one of which is indicated in the drawing at XVI. In some such cases the workers provide for only temporary support of the portable centering board, for example by light, secondary scaffolding attached to the main scaffold. This detail is not shown in the drawing.

Numerous variations of each method are known. For example, when armatures are used they may, as already noted, be of the permanent, stone-constructed type. In that case Figure 7-I should be interpreted as showing a vertical section through a part of the stone rib and its own temporary armature. The wall masonry may then comprise only one or a few of the courses of stones, shown in the left-hand side of the illustrated area. The wooden armature is removed when the stone rib or permanent armature has been built. The wall masonry, in process of construction, can be supported, for example, by stone-weighted ropes tied to temporary scaffolds on the stone ribs.[4]

When using freehand vaulting, the workmen must properly position their portable centerings and the secondary scaffolds for them. They must determine the proper positions by suitable measurement. Various measuring systems, not shown in our drawing, are known for this purpose. (In modern

[4] Gothic rope centerings are shown by Fitchen, *The Construction*, p. 182, while portable centering boards are shown *ibid.*, pp. 117 ff. Discussion: *ibid.*, pp. 183 ff.; L. R. Shelby, review, *Technology and Culture*, II, 1961, pp. 400 ff.

times, the workmen can, for example, use some telescopic instrument such as an "engineer's transit.") The principal measuring device, available since ancient times, is a plain wire. It must have one end attached to the internal scaffold, below and to the right of the area shown in 7-II. The wire must have a fixed length, equal to the radius of the desired vault. Its free end can then be used directly to position the portable centering shown in the drawing.

As an alternative procedure, it is possible first to use a wire or a rope to position a first portable centering, for example a vertical board, shaped to outline the corner of an octagonal vault. It is then possible to use the edges of such corner centerings as guide lines for positioning other wires, ropes, or centering boards that can extend in a horizontal line from one corner to the next. Ropes themselves can also be used as centerings. They can be attached to an outer scaffold provided for this purpose, run over the last course of stones, and have a free, weighted end hanging into the cupola.

Other and minor variations are also possible for the cupola masonry itself. The top edge of this masonry can be kept flat, as in 7-I and 7-II, for maximum convenience of the bricklayers and stonemasons, or this edge can follow various slopes. Within the masonry, various patterns can be used, including plain and uniform courses as shown, or a herringbone pattern, *spinapescie*, which was well known in various ancient and medieval traditions. A Florentine application of a herringbone pattern, in a plain semicircular vault, is shown in a drawing by San Gallo, but, as noted by Vassari, such pattern is not directly or easily applicable to a polygonal vault.[5] Further problems arise of course in the case of concentric cupolas.

What vaulting methods were in vogue in Florence before Brunelleschi? We doubt that the most developed Gothic methods were known to such men as Giovanni d'Ambrogio. More may have been known to Ghini, since he was closer to the active phase of Gothic construction. Arnolfo may have undertaken the project in mere reliance on some future, God-sent master, who would find ways and means to cover the large octagon and who would then develop his own forms in accordance with such means. This attitude was widespread in medieval construction work. In any case, no one in any country had ever built or designed an eight-cornered armature of the heroic size required by the plans of Arnolfo, Ghini, and the artists. Even today, after Palladio and his development of truss work, it would be extremely difficult to build such a wooden monster. We think Vasari's view is correct: no one could then build a rigid, wooden armature as required by the plan as inherited, not even by spending unlimited sums of money on it.

[5] The drawing (No. 900A^v, Uffizi, attributed to Antonio da San Gallo the Younger, reproduced in *Sanpaolesi*, Pl. 31) has a text: "Round brick vaults, as built without armature in Florence," *volte tonde di mezzano quali si costruiscono sanza armadura in Firenze*. Compare Vasari, p. 205 (a fictitious speech), about the difficulties encountered except in case of the "round vaults."

Nor could the Florentines, so far as their literature shows, successfully build a *cintre permanent* of the required size, as this required armatures for arcs that far surpassed the local experience. However, it is true that bridges of comparable span had been built elsewhere.[6]

A new idea does not come from a vacuum. The freehand vaulting method may be seen as a development of the Gothic *cintre permanent* method. This can be explained as follows. As the work according to 7-II progresses upward, the uppermost work surface (flat or inclined) rises with it. This surface is only a passing element of the ultimate masonry structure. The ultimate thrust of the cupola reacts against internal surfaces, following the heavy zigzag line in 7-II, which is approximately normal to the masonry surfaces facing in and out. This heavy line may be considered the physical work surface, while the uppermost zigzag denotes the temporary, technical work surface. The worker, in depositing new material on the technical work surface, utilizes the physical work surface, preexisting, as a kind of permanent armature or centering.

A major advantage of this method is that it makes the vault construction safer, or at least makes it simpler to make it safe. In the other method, shown at I, accurate maintenance of the wooden armature XII for centering XI was a serious problem, and even the accurate removal of the centering, after completion of the cupola, was ex-

tremely difficult, due to the conflicting characteristics and limitations of wood and stone with respect to the flexibility of each material. Even when steel is used in modern times, much effort and ingenuity are needed for proper adjustments in the maintenance and removal of a centering and its armature, causing and allowing the stonework to flex only to the extent of its own, safe limitations.

We have considered methods that "can" be used to vault a cupola, and methods used by Filippo's predecessors. We come to the question: What method or methods did Filippo use in building the Cupola?

As clearly indicated in his specification, point 12, he omitted armatures and used an unconventional scaffold. The use of inner armatures and outer scaffolds was conventional, and we conclude that he used, instead, an inner scaffold, that is, a device of the type shown in Figures 7-II and 8. As shown by the Opera documents, provision was made for eight centerings of some kind, without armatures, perhaps one centering for each corner or side. In addition, Filippo used a system of three measuring wires.[7] We do not know whether this was his complete instrumentation, and since we have no authentic illustration of the centerings and wires, and very little of the scaffold, we prefer not to speculate about

[6] Strauch, *History of Civil Engineering*, pp. 49 f.

[7] *Cupola*, Doc. 170 f. (1420), *centine;* Doc. 75 (1425–1426), *gualandrino.* The latter document (printed in our Document section) may indicate that the *centine* of 1420, as well as all conventional centerings, were omitted, but this is inherently improbable and not supported by other evidence.

the details. As to other techniques, we notice from visible parts of the Cupola that he often adjusted the brick patterns to maintain a flat working surface. He made use of herringbone patterns, but only occasionally as indeed was necessary in view of the octagonal design. We conclude that he used the freehand vaulting method schematically illustrated in Figure 7-II; however, we can not be sure of the form of his centerings.

We think he invented this freehand vaulting. There is no record of earlier uses or proposals of this method. Filippo developed freehand vaulting at various places, even before he built the Cupola of Santa Maria del Fiore. Manetti refers to his construction of a small, early cupola, with internal ribs, *a creste e vele*, and mentions that he used

> . . . a cane or measuring rod [not yet the measuring wires], fixed at the lower end but the entire body of which was able gradually to rotate upwards, this cane always restraining and touching with its free end [probably with a board at this end] the bricks that were in process of being placed, until the vault was closed.[8]

The old humanist's report is technically vague but is fairly interesting. It shows that Filippo tried successive forms of portable centerings, without conventional armature, in the con-

struction of different vaults.

We have graphic evidence for still other variants and applications of the freehand method.[9] This evidence is found in a so-called *castello* or scaffold picture (Figure 8), published long after Filippo's time, in a book by the later Opera superintendent, Nelli. It may be connected, in part, with the illustration of a similar *castello* used by Della Porta in the vaulting of the cupola of St. Peter's basilica.[10] The illustrations are interesting, as they contain evidence (not easy to find elsewhere) that freehand vaulting, Brunelleschi style, continued to be used in a major building enterprise.[11] Nelli's drawing may also be interpreted as showing that Filippo or others proposed, and perhaps used, his method also for smaller Florentine cupolas vaulted after Santa Maria del Fiore. The drawing shows, peripherally, the characteristic outline of such cupolas as those of Santo Spirito and the Pazzi Chapel. It does not, as previously assumed, show a somewhat distorted outline of the Duomo's Cupola and Tambour. Nelli asserts in a text appended to the drawing that an original of the drawing came from Filippo's own hand. The interpretation of this assertion is not clear, but it seems probable that Filippo at least indirectly guided the hand of a draftsman who

[8] *Manetti*, p. 37. Also see G. Scaglia, "Drawings of Machines for Architecture from the Early Quattrocento in Italy," *Journal of the Society of Architectural Historians*, XXV, 1966, pp. 99 ff. (hereafter: *J.S.A.H.*).

[9] The evidence was overlooked by Prager in *Osiris*, IX, p. 461 and *passim*, and by others.
[10] Fontana, *Templum vaticanum*, p. 321; also see his text, pp. 480 f.
[11] Prager, in *Osiris*, IX, p. 506, assumed that the method did not continue to be used at all. The other literature is silent about this question.

produced the original drawing for Nelli's printed illustration.

For the onlookers, these developments were as complex as they were astounding, and such a person reacted to them by merely quoting the words "vaulting without armature." Even Alberti in his early work, *Della pittura*, merely repeated this formula, in substance, and in his main work, *The Architettura*, he was silent about vaulting. Did he understand Filippo's method? Approve it? Compare it with Roman methods? As his prototype, the work of Vitruvius, was silent about vaulting, he had nothing to add, not even about this method of unequaled fame. The difficulty may have been due in part to the lack of proper terms. An invention cannot be described usefully by merely saying that it works "without" a conventional instrument. The positive terms used for Filippo's scaffolds, centerings, and other pertinent devices were vague. Thus the method became, in substance, a legend.

In fact, people came to use the phrase "without armature" in the widest and most indefinite sense.[12] They attempted to explain the meaning of the phrase in the most contradictory ways. Some considered the reported method a thing so extraordinary as to be incredible, while others glibly asserted that it was old and obvious.[13] It has been said that

the method merely was one of Filippo's practical jokes. One modern historian considered the method incapable of application in the upper layers of the cupola. Another believes vaulting without centering or armature relies, in essence, on a use of herringbone masonry.[14] Only one historian sensed that a specific technique was implied—the use of small, segmental centerings.[15] No one seemed to note even the simple, basic fact that the method used inner scaffolding instead of inner armatures, although the scaffold drawing published by Nelli was widely known.

Some further words must be added about the history and significance of vaulting without armature. In Florence the method was highly valued, as it allowed, among other things, large savings of lumber, a material in extremely short supply.[16] Builders had often explored possibilities of wood-conserving methods but had not achieved anything that could compare with Filippo's method. Admittedly, this point is in need of more detailed study, as many comments in the existing literature are all too confused. Some of the confusion is due to the fact that Ghini, the *capomaestro* in Ser Brunellesco's time, had received recognition for an invention, said to relate to "the

12 *Vasari*, p. 223, seems to use this term, in error, when he (following *Manetti*, p. 56) describes a building project on which Filippo had worked without a model.
13 A Choisy, *L'art de batir chez les Romains*, Paris, 1873, *passim*. An opposing view is voiced by G. Giovannone, *La tecnica della costruzione presso i romani*, Rome, ca. 1925, pp. 21 ff., 37.

14 G. Uzielli, *La vita e i tempi di Paolo dal Pozzo Toscanelli*, Rome, 1893, pp. 45 ff.; *Fabriczy*, pp. 80, 86, 94 f.; *Sanpaolesi*, pp. 58 ff. and Pls, 18, 28, 30, 32, 53.
15 D. M. Manni, *De florentinis inventis*, Ferrara, 1731, cap. XLII: "With no support-centering or with a very small one" (*absque vel minimo substentaculo*).
16 *Costruzione*, Doc. 64, and Doc. 70 at pp. 70, 86; V. Crispolti, *Il Duomo di Firenze*, Turin, 1938, p. 248.

armature to be used in connection with the arch of the Cupola."[17] This "arch of the Cupola" has sometimes been identified with the Cupola itself. Conceivably the expression means that Ghini planned to use the old *cintre permanent* method. There has been confusion between this reported armature and an alleged, early idea about dispensing with armatures.[18] To us it appears most probable that Ghini had received recognition for some improvement, now forgotten, in an otherwise conventional armature. This has little to do with Filippo's freehand vaulting.

Later books on architecture paid little or no attention to Filippo's method or to freehand vaulting in general.[19] It is almost certain that practitioners in Florence and Rome continued to use Filippo's method for many decades, but writers seemed to disregard it. Nor is the method widely used today. The modern "Thin Shell" vaults are generally constructed on vault-wide armatures, formed of steel trusses and supporting centerings of wood or wire or fabric or combinations of various materials, whereon the masonry is poured as a thin layer of concrete. Some modern builders use improvements inconceivable in Filippo's time, for example, a centering consisting of a balloon inflated by compressed air, with automatic pressure regulation to compensate for the growing weight of the concrete shell.[20]

It appears that the sustained, literary fame of Filippo's method may be . greater than the ultimate, technical significance of this ingenious invention. Perhaps it was its main historic importance that it helped psychologically to merge medieval and rediscovered classic methods as it demonstrated the possibility of liberating the arts from assumed limitations. At least for a time it overcame a limitation dictated by rules of long standing, at least in Florence.

CONFLUENCE OF CLASSIC AND GOTHIC

In 1367, the Florentine artists had decided to dispense with Gothic buttresses and had expressed hope, mixed with some well-founded anxiety, that it would be found possible also to dispense with visible tie rods. In the next generation such men as Giovanni d'Ambrogio wished to return to the buttressing system. The invention of a stable structure without buttresses for a

---

[17] *Costruzione*, Doc. 231.

[18] Speculations about Ghini's armature were announced in various works, including the first edition of this study. As Saalman, "Santa Maria del Fiore," p. 492, plausibly writes, we may never know what Ghini's invention was. However, according to a personal communication of Dr. Saalman, Ghini's armature (for the Cupola) was part of the artist's model that he built, and it was saved when that model was destroyed in 1421.

[19] The few remarks about vaulting in Alberti's *De re aedificatoria*, Florence, 1485, VII, xi, 126 f., presuppose a full-sized *armamentum* as a matter of course.

[20] For general orientation: C. W. Condit, *American Building Materials and Techniques*, Chicago, 1968, pp. 275–278. For some of the details see for example J. Joedicke, *Shell Architecture*, New York, 1963; D. P. Billington, *Thin Shell Concrete Structures*, 1965. Pertinent patents are in U.S. Patent Class 25–131.

large cupola was a problem of the century, and Filippo solved it.

He solved it by his new combination of walls with a new form of Gothic ribs and Classic tie rings, which he then developed further by his use of additional reinforcing arch elements. Recognition of these inventions figures prominently in the second of the prize awards that he received, a document of 27 August 1423.[21] He received the prize for

> . . . designing the model of the chain to be installed in the main Cupola and bringing it to perfection, and for several other devices made by him in said Opera, that is, devices to be made yet, such as finding the way how windows shall be placed in the Cupola, how the stone chain shall be arranged, and how the ridges shall be placed on the Cupola.

Evidently this was a notary's description of Filippo's Cupola work, as of the year 1423. It is technically naïve and groping, in contrast to Filippo's own specification written three years before, which had answered most of the questions suggested here and had gone far beyond these generalities. The official document shows only that several points of the existing program and one newly added point, windows, were under debate; it does not state what "devices" Filippo had designed. In some way a chain was still believed to be in need of "perfection" and some devices were still to be made, although Filippo was recognized as the originator of the structures "to be installed." It appears that, in principle, the elements

21 *Cupola,* Doc. 177.

specified in 1420 were approved and, in fact, now were thoroughly appreciated, as shown by the prize award. We shall briefly review these elements once more, to show their relation to this award and to Filippo's historic fame as originator of structural principles.

Two concentric vaults interconnected by ribs in vertical planes were the fundamental structure. These internal ribs can be seen as Gothic buttresses of the inner vault and at the same time as Gothic ribs of the outer vault. By 1423, they were no longer controversial. The prize document mentions only the outer, visible ridges of these ribs. Evidently the Florentines began to see how Filippo proposed to give visual expression to his structural concepts, even if they hardly undertook to understand these concepts in a technical sense.

The prize document gives, at most, some nonspecific recognition to Filippo's use of various horizontal tie and reinforcing members, which he had specified in 1420. One group of these members comprises the "small barrel vaults" that connect the corner ribs to adjacent intermediate ribs in the hollow space between the shells. The document speaks only of unspecified "devices," other than the stone chains and the ridges. Somewhat later, one set of these small vaults was discussed (if we are correct in so interpreting a document of 1425–1426). The placement of the vault had to be reconciled with the arrangement for walkways and doors in the hollow space. No real controversy seemed to result. The "small vaults"—then also identified by another term—were mentioned in a way that apparently caused no public debate and no interest on the part of the biographers. It appears that no one in the councils of the Opera cared to discuss them and that only a few of the builders understood their structure and function. Meanwhile, by contrast, there was some discussion of the Classic tie rings, in the prize award of 1423. We must turn to the first of these rings.[22]

THE WOOD CHAIN

Filippo's wood chain is described in the second part of Point 7 in the specification of 1420.[23] This chain is a single ring of timbers interconnected by iron bolts, secured to the masonry by iron chain rods a short distance above the Tambour. The ring passes through each rib, and between the ribs it extends freely across parts of the hollow space. If there were any strain (a minute beginning of motion) of lower stone blocks or brick masonry, due to the weight of the upper cupola layers, this would at once cause tensional stress, and such stress would also occur in the wood chain. This stress would apply bending and shearing forces to the bolts interconnecting the timbers or the chain rods anchoring them to the masonry.

The wood chain was not only specified by Filippo but also, in part, provided by him as architect-contractor. It consists of twenty-four massive logs, each about twenty feet long and having a cross-sectional size more than a foot high and wide. The logs are of chestnut, not of oak as originally specified, a change that was made at the end of 1421. The logs are interconnected by rectangular "fish-plates" of oak, two at each joint, and each about one-half as thick as the logs. The connecting plates are fastened to the logs by square-headed iron bolts. There are iron straps wound around the joints and nailed to them, obviously to prevent the bolts from splitting the wood when the chain is under tension.[24]

The devices connecting the chain to the masonry, not previously described in the literature, are found near some of the ribs, where the chain enters the stone structure. At these points Filippo provided iron rods or wall clamps (also called "chains," "chain rods,"

[22] The horizontal ties are generally called "rings," *cerchi*, for example, in points 5 and 6 of the specification, but the wooden tie is called a "chain," *catena*, in point 7; also see Docs. 173, 193: *catena de' castagni*. As explained in notes to these points, the word "chain" also has other meanings. The documents also distinguish between *ligamen* and *lignamen*: Doc. 177 mentions the *chatena . . . ligaminis* or tie chain, of stone and iron, while Doc. 175 speaks of its *modello lignaminis*, or wood model.
[23] It is illustrated by *Nelli*, Figure VIII A, whose drawing is copied and amplified here as Figure 10, parts II and V. The payments were made near the end of 1423 (*Cupola*, Docs. 188 ff.). No earlier ring-shaped wood chains are known to us. Vitruvius, I, v, 3, had provided for "ties made of charred olive wood, binding the two faces of the wall together."

[24] *Cupola*, Doc. 188: a loan of 10 florins, granted to Filippo for this contract on 27 August 1423. The iron straps were repaired in 1825: *Cupola*, p. 195.

catenae or *catene*). Such a rod, a well-known building element used since time immemorial, has a loop or "eye" at each end, engaging a strong iron cross rod. In Filippo's Cupola these cross rods pass through the wood chain structure. The cross rods are tightened to the iron loops by wedge blocks, tightly driven into the loops behind the cross rods (Figure 10, Detail V).

It is not authentically recorded why Filippo used the wood chain in addition to his stone rings, and conflicting answers have been proposed. It has been said that this was merely a compromise with methods of the past, which had used various types of chains in smaller vaults. This answer disregards the specific way in which Filippo describes this chain in connection with the small barrel vault elements (Point 7 of the specification). Others say the device was of little structural utility,[25] or was even useless. Some writers say it was one of Filippo's tricks to mislead Ghiberti (but he must have convinced the Opera and its building masters to approve payment for it). Still others feel that the chain withstood some significant part of the crown thrust existing in the Cupola. (This point, however, could be demonstrated only if more facts than we actually have were known about the anatomy of the vaults). In the absence of authentic and persuasive answers, it becomes necessary, as at other points of the record,

[25] G. B. Nelli, *Discorsi di architettura*, ed. G. B. C. Nelli, Florence, 1753. According to Nelli's son (biographic introduction to this work), Carlo Fontana disagreed. Also see *Cupola*, pp. 209 f.

to determine what the structure "could" achieve.

The static outward pressures or "crown thrusts," caused by the weight of the Cupola materials and present in their mass, impose a "hoop tension" on circular layers and chain systems in this mass. This tension is counteracted and absorbed by the stone and wood rings that Filippo installed; the masonry itself has hardly any tensional strength. The wood chain contributes toward the required absorption of this load, although this chain probably contributes only in limited measure to the effect provided by the more numerous, more substantial, and totally embedded stone rings.

The joints of the wood chain are in exposed position, not within the wall masonry as the stone chains. This fact may be a clue to a function of the wood chain. The exposed joints could indicate the development of thrust forces in the Cupola, and the wood chain could, accordingly, be used as a gigantic testing device. At some 24 points one could observe any relative motion of the logs, by determining how firmly the wedges were held in the joints.

Of course it was not easy to interpret such observations, if indeed it was intended to make them. The reason is that wood logs are also subject to shrinkage and expansion in response to changes of their humidity and temperature. In the normal cycle of the seasons a ten-foot length of a wood beam could expand and shrink about one-eighth of an inch, which in turn could cause deflection of the beam as well as changes of the force pattern at the joints.

Filippo's attention to the exact materials used in the chain indicates that he may have observed such *minutiae*. Therefore we think he may have used the chain as a testing device, while also using it more basically and conventionally as a tie element. The construction is fairly similar to the test means used in later centuries, although it is enormously larger. In earlier times, as we have noted, the builders observed the development of hairline cracks. In and after 1379, Blasio del Abaco made measurements of the octagon foundations. In later times the successors of Filippo made more systematic tests, when they studied the statics of large cupolas. Perhaps the wood chain of the Florentine Cupola was a way station on the road from these earliest beginnings to modern, scientific analysis of stresses and strains.[26]

Conceivably the wood chain also served to avoid dangers connected with the working of the hoisting machine. When this machine raised a heavy load, the legs of the scaffold supporting the upper wheel for the hoisting rope must have vibrated strongly. If the principal scaffold drawing (Figure 8) is interpreted as representing the Duomo structure, it may indicate a construction for channeling such vibrations into the wood chain, as the drawing shows beams extending from the legs of the scaffold toward the vicinity of this chain.

In any event it appears that the workmen were impressed and encouraged when this chain had been installed, near the end of 1423. Perhaps we should leave the last word about this strange development to the biographer.[27] He describes the ties of the masonry, including the wood chain, and then says, "The ties were now finished right round the eight sides. The masons, being encouraged, were laboring valiantly." Either they felt the masonry would be safer, or at least its safety would be surveyed more closely, as the wood chain between the lower and middle levels of stone chains went into place.

THE STONE CHAINS

The Cupola has chains or so-called rings of sandstone blocks interconnected by metal clamps and tying the walls and their ribs against outward spreading, according to points 5 and 6 of Filippo's specification. These chains are visible only in small part, where surrounding masonry has been omitted or removed or cracked. The basic arrangement of the chain rings is shown

---

[26] See *Costruzione*, Doc. 176, for the observations of 1366, and Docs. 302 ff. for those of Blasio. See *Cupola*, Docs. 382 ff., 391 for observations made in the 1600s and early 1700s. For more recent observations: Opera di S. Maria del Fiore, *Rilievi e studi sulla cupola del Brunelleschi*, Florence, 1939, pp. 18–24 (hereafter: *Rilievi*). Also see J. P. Richter, *The Literary Works of Leonardo da Vinci*, II, New York, 1939, pp. 77 ff. For the history of civil engineering the most important observations of the Florentine Cupola were those of Manni and Poleni, see Poleni, *Memorie istoriche*, II, xviii.

[27] *Vasari*, p. 218.

in Figures 9 and 10. Although the structure has been studied by Nelli (ca. 1700) and by Nervi and others (ca. 1934), it is not nearly as well known as we would wish. It may be possible to locate the stone blocks in the masonry, and even the metal clamps on the blocks, by methods of nondestructive testing, relying on echo analysis; but until such methods are approved, performed, and evaluated, we are largely dependent on the mere information contained in old documents and in present-day features visible at the surface.

The rejection of Gothic buttresses and the hope for good and unobjectionable inner ties had been cardinal points in the debates of 1367. From then on the question was how to design these inner ties, if possible, without extending them across the vault. The paragraphs about the tie chain in the specification of 1420 gave the first explicit answer to this question. The document provided for three pairs of tie rings composed of stone blocks and clamps, aside from the wood chain. The specification did not indicate the levels where the upper stone rings were to be installed, although it made it clear that the lower rings were to be incorporated in the bottom of the Cupola. The model may have shown the elevations. In the structure itself a ring of small windows illuminates and vents each inner walk, and these walks rest on radial beams, which in turn are connected with the tie rings. Was this construction new to the world, as it apparently was to Florence?

An elementary form of the tension rod had, of course, been used since time immemorial; the horizontal tie beam of a plain roof (the *monaco*) is such a tie rod or tension rod. It prevents the rafters of the roof and the weight thereon from pushing the side walls of the building apart. Antiquity was well versed in the art of constructing such ties of wood or metal, interconnected by suitable dowels or clamps.[28] Ancient builders also employed peripheral ties in circular and polygonal vaults, either exposing them to sight or, more often, embedding them in the masonry. When Roman architecture was at its height, it made regular use of such tie-ring systems, using metal clamps. In some cases the rings were embedded in wall structures reinforced by radial ribs. It appears that Brunelleschi rediscovered this feature, a rediscovery that may be called one of Filippo's major contributions to architecture. It should further be noted that his structure goes beyond the principles that had been used in the past.

If Filippo's Cupola were limited to its inner shell and so much of the rib and tie system as is embedded therein, the structure would be a cupola of Roman type. By contrast, if it consisted of its inner or outer shell without tie rings but with the projecting ribs or but-

[28] For a most ancient reference, see II *Chronicles*, 34, 11. Also see Giovannone, *La tecnica*, pp. 19 ff., 52 ff. and *passim;* S. B. Hamilton, "The Structural Use of Iron in Antiquity," *Transactions of the Newcomen Society*, XXXI, 1957–1959, pp. 29 ff.; Strauch, *History of Civil Engineering*, pp. 10 ff. We understand from Professor Millon that the C-clamps of iron, known from antiquity, still appear in the cupola of San Pietro. We conclude that they probably are used in Brunelleschi's Cupola.

tresses, the structure would be in essence a Gothic vault. In either case it would have less stability than it gained by its actual cellular and peripherally tied construction. It appears to us that it was a new invention, combining Roman and Gothic features.

Gothic builders had made substantially no use of tie rings. If some builder of that epoch used such ties in one of the rare, octagonal Gothic cupolas or *tiburio* vaults, such exceptional practice was unknown in the Gothic world at large. The combination of stone and iron members constituting such a tie ring had long been forgotten, along with other advanced techniques of the iron age at the end of Antiquity.[29] The reintroduction of metal-connected tie rings in a cellular vault fabric by Brunelleschi at the end of the Gothic age appears to us as a major part of his "renewal of Roman masonry." His renewal and development of this construction may also be seen as a long step toward modern structural engineering. To this extent, Filippo's admirers seem to us to be well justified in their appreciation of his work. He initiated a new architectural construction of greater strength than either Roman or Gothic forms had provided.

There had been partial anticipations. Others had used peripheral tie chains in small Romanesque or Gothic buildings.[30] Others may have seen the ruins of Rome. They may have speculated about ways to overcome the limitations of "ancient" vault architecture. However, Filippo became the first to make explicit and impressive use of the newer, stronger constructive form. Perhaps Neri had desired such a form, but if so, he had surely failed to describe it or construct it. Filippo described it and constructed it.

The exact times and circumstances of Filippo's discoveries and inventions are unreported, but pertinent language may be found in his biography by Manetti, where his Roman sojourn is described:

As he looked upon the sculptures [which he sought in his frustration at not being found the first of the modern sculptors] he saw the masonry, the methods of the Ancients, and their symmetries. It appeared to him that he recognized a certain order of members and supports, which became very evident, like the order given by God. He was enlightened with respect to great things, and he felt it strongly.[31]

This statement of Manetti may be derived in part from Filippo's own report, since the two were acquainted when Filippo was old and the biographer was a young man. In full accord with the Trecento documents, the biography indicates that a seed, planted in Filippo's mind during childhood, came to fruition during one of his early archaeological tours in Rome, soon after the Battistero and Duomo controversies of the early Quattrocento.

---

[29] This is true also of Italy, where Roman traditions otherwise survived longer than elsewhere. See Ward, *Medieval Church*, p. 57; Briggs, *The Architect*, pp. 93 f., 102. As a result, Gothic builders generally used only "one material, stone, which is trustworthy only in its resistance to compression" (Fitchen, *The Construction*, p. 3). Very rarely did Gothic builders use tie rods (*ibid.*, pp. 275 ff.).

[30] *Costruzione*, Doc. 70 at p. 91.
[31] *Manetti*, p. 18.

Ten years later he entered the debate that brought total success of his new conceptions.

Therefore, although competent scholars have held that Filippo's plan underwent changes after 1417, we believe that his basic and major concept, the use of a cellular framework of stone and metal, was complete by 1415, if not 1405, and that it remained unaltered in every essential part. He defended it with perfect consistency against a city full of doubters and opponents. While conducting this defense, he also introduced the new form that made him famous as originator of Renaissance forms, the *Innocenti* porch, a purely formal innovation without new concept of structural technique. It is clear that he had reached full maturity as artist, and it seems to us that he was equally mature, at that time, as an architectural engineer. His model of 1418 then showed that a vault, constructed in his new form, could stand. This showing was sufficient for the Dome administration to dismiss the then-acting architect and after a few years to accept Filippo as the new architect, although subject to conditions of a controversial character.

Promptly after Filippo's election as Dome architect, stones for the new masonry framework began to be acquired. From then on, only secondary details of this framework remained to be developed. A model of 1421 showed the exact way of tying the stones of the tie rings into the ribs.[32] No doubt

[32] See *Cupola*, Docs. 238 (30 April 1420) and 182 ff., Summer 1420, for the deliveries of stones, and Docs. 173, 185 for the model work.

similar models and developments followed in later years. They were no longer exposed to major controversy, although the Opera occasionally complained about Filippo's single-mindedness and "rancors." It seems clear that by 1418 the mind of the Florentines had been prepared, although it never ceased to be critical. We think Filippo had prepared it from 1404 to 1417.

Of course he had not solved all the problems that ensue from such a major innovation. It did not take long before some sharp-eyed Florentines realized that the light falling into the vault through the windows of the Tambour would be insufficient and that the Cupola would be dark inside. They promptly began to ask for wide-open light ports, *lumi*, in the Cupola shells, just as the contemporaries of Ghini had asked for wide windows, *occhi*, in the walls of the nave. (The proportioning of the windows for the nave had been repeated, more or less effectively, in structures designed or built in the meantime, such as San Petronio in Bologna.) It is certain that Filippo was a great friend of light and vision, and no one can have regretted it more than he that his vault is dark. However, it had fallen to him to solve one historic problem in his time. He provided light ports in his smaller cupolas, not in the Cupola of the Dome. He rejected compromise with the static principles that he had discovered and invented.

Others, less single-minded about the primary matter, kept asking for more light. Among these petitioners were two co-appointees of Filippo, the painter Pesello and the writer Acquettini. The biographers call them partisans of Ghiberti. These window advisers submitted new models of framework systems in 1424, for which they received small awards but which did not, it seems, influence the course of events.[33] As we have noted, Filippo made minor changes of his own in parts of the rib, chain, and wall structures but made none that allowed for windows in the Cupola.

One of Acquettini's petitions for windows was made in 1425. Its principal argument may be noted here:

> This demonstration of this circular aperture shows that sunlight enters opposite the piers, where it breaks and gives light by reflection. Now let any one consider whether this reflection will be of such force as to go up by a hundred-forty feet. I give notice that it seems to me necessary to provide for light before further masonry is placed. Five years and more have already passed since I [after reading the specification of 1420] gave my method for this. I make 24 windows [one between every two ribs] directly above the edge [of the Tambour] .... Thus you can see how wrong the building is, as the vault starts from this edge ....

According to the biographers, Lorenzo Ghiberti sided with this man. We have seen that his view also had a faint echo

in an early statement by Alberti.

Filippo did not allow any of the cell walls of the Cupola to be left unfilled by masonry. He had dispensed with the buttresses of the Gothic past and also with the armatures used since earliest times. Now he insisted on full use of the construction he had recognized as statically safe for the large Cupola. He developed other constructions, including several with large windows in the cupola shell, in other buildings, experimenting differently with possibilities of Gothic-Classic synthesis. Unfortunately the dates and details of the other buildings, and of their designs, are less clearly documented than those of the major Cupola.[34]

BRUNELLESCHI AS STRUCTURAL ENGINEER

The sources of Brunelleschi's ideas must be found in a Scholastic-Humanist world dominated by the Romanesque-Gothic structure of Arnolfo, Neri, and Ghini. Beyond this, little is known of Filippo's schooling, his early technical experience, and the engineers who influenced him.

It is conceivable that he had contact

[33] Cupola, Doc. 181 (April 1424), Doc. 60 (1426). Acquettini's petition is published in C. Guasti, Belle arti, Florence, 1874, pp. 115–120, 125. About Acquettini also see H. Saalman, "Giovanni di Gherardo da Prato's Designs," Journal of the Society of Architectural Historians, XVIII, 1959, pp. 11 ff.

[34] Some Florentines must have been aware of the existence, in northern countries, of Gothic vaults having large windows between narrow and widely projecting buttresses (see for example Ward, Medieval Church, pp. 128–157). Many influential architects of the Trecento and Quattrocento may have rejected these more strongly than Filippo did. We cannot find a factual basis for the views of A. Schmarsow, Gotik in der Renaissance, Stuttgart, 1921, pp. 27 ff., about either chronology or analysis of Brunelleschian designs.

with men of science, as he belonged to the higher classes of Florence. According to Manetti he knew Paolo dal Pozzo Toscanelli, a mathematician. However, nothing specific is known about talks between these men or work that they shared. If they ever discussed any of Filippo's architectural problems, the mathematician could hardly assist the practitioner in anything greater than perhaps the exact computation of some weights of material. There was no tradition and in fact not even a beginning in contemporary science whereby anybody could have solved a more difficult problem, for example, the permissible loading and required dimensions of a building element such as a column, a pier, a rib, or a cupola shell. All these were designed on the basis of the roughest estimates, traditionally based on enormous overdesigning. Nor does it appear that Filippo himself ever undertook a mathematical or analytic investigation. His biographers describe some details of his perspective studies; even here they do not mention any part of the mathematical analysis that they knew to be pertinent. Nor do they suggest any other scientific activity of Filippo, beyond his quoting Dante and the Scriptures, which they mention with naïve defensiveness. They are not even aware of experimental observations, for example, a study of the behavior of the wood chain. The only real echoes of a scientific-technical study come from the rare reports about talks with other masters.

Filippo had one such talk with Mariano Taccola in Siena, and Taccola recorded it in a note that we consider in another study in this book. The report is totally devoid of mathematics, but may suggest an interest in experimentation to explore the secrets of mechanics, hydraulics, and pneumatics. From Taccola's note and the entire evidence, it does not appear that Filippo had an urge to formulate his scientific or technical findings, particularly in written or illustrated form—although he was able to produce pungent expressions, as is shown by his sonnets and specifications. We hear some remark, without detail, about drawings that he made; they seem to be lost.[35] He worked extensively by means of models and full-scale constructions and not in general by writings or drawings.

He had pupils and admirers who witnessed his performance. Some of them in turn left a secondary record of his teachings, which in due course found a wide audience. For example, the engines in Taccola's treatises that have unmistakable Brunelleschian features reappear in many notebooks and treatises of the later Quattrocento, and so do the treatises of Taccola's indirect follower, Francesco di Giorgio.

The existence of this secondary, graphic record was overlooked in the first edition of this study, but since that time many versions of Quattrocento books and portfolios of drawings were

[35] Full-scale drawings on sand are reported by G. B. Gelli (ed. G. Mancini, *Arch. stor. italiano*, 1896, pp. 32 ff.), and such techniques were widely known (Briggs, *The Architect*, pp. 89, 128). About Filippo's Roman drawings on ruled paper or parchment and his perspective apparatus, see *Manetti*, pp. 10 f., 20 f., 56 f., 78 f. Also see p. 28, about the question of cross-sectional views in orthogonal projection.

found; a beginning has been made toward establishing their sequence and nature. The machine drawings will be discussed in a study that follows. We cannot try at this point to describe the detailed architectural contents of the notebooks. Nor can we explore their complex relationships with the Renaissance of architecture, and mainly with that of Vitruvius (who in our opinion was substantially unknown to Brunelleschi and Taccola). However, we draw attention to the existence of these architectural drawings. They are combined with the machine drawings. The drawings of Brunelleschi's own time were collected, copied, recopied, developed, confused, and improved, and the resulting knowledge was studied by later architects, until this knowledge became universal and elementary. The drawings now known, in addition to those of machines, include representations of construction methods, foundations, building tools, building elements, building layouts, and various assemblies of buildings.[36] Further studies of this record will be needed if it is to become clearer than it is today how Filippo's conceptions entered into his mind or arose in it, and how he used and developed them.

At the present point we can see with some assurance that he used and developed both Gothic and Classic principles. He was an innovator of forms that remained fundamentally Gothic, and he was just as remarkable in this respect as he was in his well-known action as renewer of Classic forms. He made distinctive although limited use of buttresses and even pinnacles (on his Lantern) but avoided the use of transverse tie rods as he developed a "masonry" strengthened by at least four reinforcements: ribs, stone rings, wood chain, and segmental arches. Here he proved himself a more original and a more successful innovator than his predecessors in the Trecento and early Quattrocento. Here he also was much more successful than later architects, who undertook to continue the construction of the cathedral of Milan.

Unfortunately part of his work remained unfinished. His followers did not complete his Cupola. They failed to build the Gallery intended to harmonize with his Exedras; yet this Gallery would be needed to give full and clear expression to his structure. Baccio tried to build it; Michelangelo halted the attempt. No one actually built it. There is evidence permitting at least partial mental reconstruction of Filippo's plan for uniting the Cupola-Tambour with the Little Tribunes,[37]

[36] Similar collections occur in earlier and later works, including, for example, the book of Villard de Honnecourt, ca. 1250 (ed. H. R. Hahnloser, Vienna, 1935); *Bellifortis* by Konrad Kyeser, ca. 1400 (ed. G. Quarg, Düsseldorf, 1967); *De ingeneis* by Mariano Taccola, 1433 (edition Prager-Scaglia, in preparation); and the *Trattati* of Francesco di Giorgio and followers, ca. 1480–1550. Also see Frankl, *The Gothic*, p. 145, about an unpublished "Vienna Sketchbook."

[37] Filippo, as well as his Trecento predecessors, may have expected to surround the foot of the Cupola with statues or perhaps with pinnacles similar to those on his Lantern. He provided platforms, perhaps for statues, in the masonry itself, at the foot ends of the outer ribs or *creste*, in or directly above the Gallery zone over-

but the evidence is not sufficiently known and analyzed to justify more than these general suggestions at present.

In its body as well as its Gallery area, the Cupola is less than perfect. It has four plainly visible although minor cracks, which extend through four of the eight sides and also through large parts of the understructure. They have been carefully studied by experts, including G. B. Nelli about 1695 and P. L. Nervi and others about 1934.[38] It

---

lying the Tambour. At each of these ends, four chain rods with outer ring-shaped ends extend from the masonry (see Figure 10, view IV, detail "2"; *Forma e colore*, Pl. 5; Alinari Photo 58042).

[38] *Cupola*, pp. 209 f., and Doc. 340 (1561). The cracks, oddly, are located in the sides overlying the piers, not in the sides overlying the large Gothic arches between the piers. As might be expected, each crack extends between the corner areas reinforced by the *volticciuole*. The facts, determined by P. L. Nervi and others and ably presented in *Rilievi*, may be interpreted as indicating that the building is settling somewhat more on its south side than at the north; that as a result the main piers and overlying masonry layers undergo shearing strains and that strong ties (perhaps in the Tambour, Figure 5) reinforce the sides overlying the Gothic arches. It is possible that earthquakes contribute to the settlement of the supports and the formation of the cracks, in a gradual way. Some five to ten earthquakes of significant strength occurred in Florence, between 1430 and 1580; fewer and lesser disturbances followed, between 1580 and 1690 (M. Baratta, *I terremoti in Italia*, Turin, 1901, p. 743). It seems that during and after the latter years the cracks were substantially enlarged, giving rise to consultations of Nelli, Nervi, and others, also in 1954 and later. A new analysis by R. J. Mainstone is about to appear in the *Transactions of the Newcomen Society*, for 1971.

was found that insertion of new tie rings would be possible but that they were not needed by the conditions thus far noted. These conditions are under more or less constant review.

The forces active in a cupola are complex. Their distribution and magnitude were unknown to Roman, Romanesque, Gothic, and Renaissance builders. Galileo, two centuries after Filippo, began to raise fundamental questions about such forces, and the next few generations of pupils, including Nelli, considered various aspects of these questions. In this way they continued the attempts of the Trecento masters and of Filippo himself. Directly after Nelli's time, while observations and studies were still limited to the cupolas of Florence and Rome, came the breakthrough of structural analysis, and the first few of the fundamental laws, now recognized in this field, were formulated.

It has been usual to say that Vasari overestimated Filippo, but it is possible that he did not estimate him highly enough. To us Filippo the architect appears as one of the great developers of Gothic building, as the principal founder of the Renaissance, and also as an important forerunner of modern structural design and analysis.

DRAWINGS OF BRUNELLESCHIAN MACHINES[1]

4 GHIBERTI's *Zibaldone*. A SECRET HOIST
There are many engineering note-books containing material of Filippo's time and of the decades and centuries that followed. Almost all show mechanical constructions traceable to the Florentine master, but such constructions are shown in particular concentration in the so-called *Zibaldone* or Notebook of Buonaccorso Ghiberti (1451–1516), a thick tome in octavo format owned by the Biblioteca Nazionale in Florence. The volume contains drawings of machinery, sculpture, and architecture, interspersed with workshop notes of the Ghiberti bronze foundry. We will disregard the sculptures and architectures at this point but reproduce some of the machinery drawings, especially some of those found on folios 94 to 120 of the book. They show devices that belong to Filippo's circle, according to various items of evidence. They deserve close review because they, together with other documents, constitute a record of unequaled comprehensiveness, aiding us greatly in our attempt to visualize Brunelleschi's methods of work.

The *Zibaldone* has notes in a Quattrocento hand, probably that of Buonaccorso, and has drawings that are, at least in great part, likely to be his copies derived from work produced or collected by his grandfather Lorenzo Ghiberti.[2] The interpretation of the drawings does not depend on this detail, but it is tempting to call the book a Ghibertian record. As such it would be entitled to most careful consideration as perhaps a direct reflection of the Brunelleschian record.

There are close parallels between this book, the official records of the Opera, and the biographies of Filippo. The book shows, as we will see, what the documents call Filippo's "edifice for pulling and hauling" and what Manetti calls his "machines for carrying, lifting and pulling."[3] It is possible that Vasari knew this book. His friend Cosimo Bartoli inherited it, as a note on its first folio indicates. Later the book became known to historians of military technology.[4] The present study is the first to consider Brunelleschian machines on the basis of this and related notebooks.

Buonaccorso's books, like many drawing collections of the Quattrocento, show a number of simple, unitary crane hoists, including an elevated one, operated by a treadwheel (Figure 12), which may be considered representative of the hoists used in the construction

[1] Studies 4 and 5 are a revised edition of G. Scaglia, "Drawings of Brunelleschi's Mechanical Inventions for the Construction of the Cupola," *Marsyas*, X, 1960–1961, pp. 41 ff. (hereafter: *Marsyas*).

[2] Early opinions about authorship are cited in R. Corwegh, "Der Verfasser des kleinen Kodex Ghiberti," *Mitteilungen des Kunsthistorischen Instituts zu Florenz*, I, 1910, p. 159 f. See now, *Marsyas*, X, p. 73; T. Krautheimer-Hess, "More Ghibertiana," *Art Bulletin*, XLVI, 1964, pp. 307 ff.
[3] *Cupola*, Docs. 123 f., 146 f. The document of 1420–1421 speaks of engines *pro trahendo et conducendo; Manetti*, p. 19, writes, *da portare e da levare e da tirare; Vasari*, p. 218, is less specific.
[4] See, for example, M. Jähns, *Geschichte der Kriegswissenschaften*, I, Munich, 1889, pp. 285 f.

12. Conventional elevated crane in Ghiberti's *Zibaldone*. Facsimile from BR 228, fo. 94ʳ in Biblioteca Nazionale, Florence.

13. Conventional ground-supported cranes in the *Zibaldone*. Facsimile from BR 228, fo. 95ᵛ.

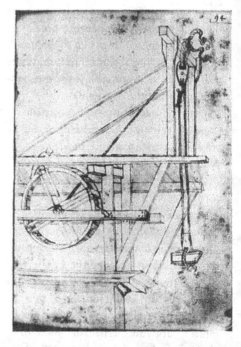

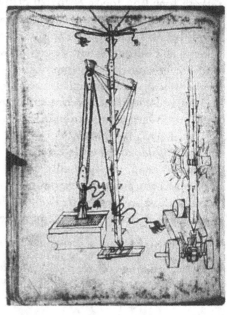

of Gothic cathedrals. In the *Zibaldone* this hoist is shown as lifting a cornice block, probably for the completion of a cornice, shown in part, the finished elements of which already contribute to the support of the crane and its scaffolding. The draftsman does not attempt to show the scaffolding entirely.[5]

Another page (Figure 13) shows ground-supported hoists, either stationary or carriage-mounted. Such a hoist has a rigid upright mast, with a boom pivoted to its top and a rope pulley at the end of the boom; the mast is arranged for easy climbing. Similar devices had been used in Antiquity, and builders continued to use them.[6] We may safely assume that such cranes were known to the Trecento builders in Florence. However, these simple cranes reached only such height as was attainable by a wooden mast, the height of which was limited by the natural height of a tree and the possibilities of bundling trees or logs. We do not know how the Tambour was constructed, but it seems probable that Giovanni d'Ambrogio used the crane of Figure 12 rather than those of Figure 13.

Ancient and medieval builders were also familiar with hoisting systems wherein a long rope was pulled by a ground-supported winch, and the rope then ran to and over an elevated pulley-wheel installation.[7] The Florentine documents show that Brunelleschi used such a system, and the *Zibaldone* shows several variants of the machine elements or components forming part of such a system. Even if it shows these elements intermixed with cannons, foundry bills, architectural drafts, and other things irrelevant to the machinery record, it provides lively and clear illustration of the devices, in a form easily understood by any one who has some familiarity with ancient or medieval machines or with the requirements of machinery in general.

Figures 14 and 15 show one of the ground installations or winch machines. These two drawings of the *Zibaldone* appear to be of secret character, as the text on these sheets is in cryptogram. Some one apparently wanted to have a record of the machine but to withhold the explanations from others. There is no pertinent information in the biographies or Opera documents or in subsequent copywork, and there is very little explanation in the cryptogram. However, the drawings are technically self-explanatory in most respects.

Although the two drawings are separated by several pages that illustrate different things, the drawings are clearly related to each other. They

[5] This crane is also known from a miniature in a codex of ca. 1300. H. Naumann, ed., *Die Minnesinger in Bildern der Manessischen Handschrift*, Leipzig, ca. 1933.
[6] Vitruvius, X, ii, 1–10; Roman relief in the Lateran museum (Strauch, *History of Civil Engineering*, p. 26; A. G. Drachmann, "A Note on Ancient Cranes," C. Singer et al., *A History of Technology*, II, Oxford, 1956, pp. 658–662, Fig. 603). Such cranes are still used, although with modern motors instead of treadwheels. Also see Wittkower, *La cupola*, Pls. 37, 39a, 41.

[7] Schematically shown in a relief at Capua (Drachmann, "A Note . . . ," p. 659, Fig. 578, with erroneous legend). They also occur in miniatures. Singer, *History*, Fig. 580, Pl. 30B, 31A.

14. "Secret hoist," from *Zibaldone*.
Facsimile from BR 228, fo. 95ʳ.

15. Same hoist. *Ibid.*, fo. 98ʳ.

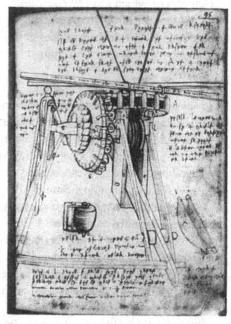

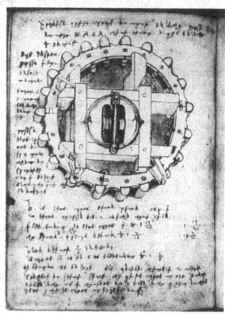

provide, respectively, side and top views of a single machine. The views are rendered in realistic form, with adequate use of line perspective (indicating that they can hardly be earlier than Brunelleschi's time), and with good characterization of technical forms and materials. The text is written in a code that replaces each letter by the preceding letter of the alphabet, for example, b by a. When decoded and translated, the description begins,

(Fo. 95ʳ) The one of these ropes is wound up and the other unwound, inside the wheel with teeth. Here in the center are two wheels over which they run. One set is below, and above it are two more, smaller, where the ropes come together, so that there are 4 wheels in wheel A, two upper and two lower ones, where the ropes run . . . .[8]

Evidently, someone labors hard to describe what to others, even then, was rather simple. There is little of any technical interest in this "secret" text, which totally misses the more interesting parts of the machine and indicates confusion about the obvious parts. It says of one gear that "its total diameter is 1⅝ *braccia* and thus it winds up one inch less than one *braccio*." It is a futile attempt to say something of a quantitative nature. It says repeatedly that inside bearing A are "four pulleys . . . the wheel is placed above the four pulleys which are on the pedestal, that

is, above the large rope-driving pulley." Just what is in bearing A is at least as "secret" after the decoding of this information as it was before. Equally confusing are many other expressions in the cryptogram, including, for example, a reference to "woodwork above the pulleys, similar to two strips (*lestoli*)." The text does not say whether the beams, fastened to the top wheel in the first drawing, are turned by an ox, a donkey, a group of men, or a single man. Thus the approximate size and power of the machine remain secret. Assuming that ratchet C is arranged for convenient access by a man (who would try to control it, strangely enough, by a hand-held tool), the drawings may be interpreted as indicating that the machine is a few feet high. It then probably weighs a few hundred pounds. An attempt to hoist a ton or more with it would result in its destruction.

The machine is unusual.[9] Perhaps its principal element of novelty may be found in the use of a large, hollow bearing, which allows the hoisting ropes to run directly upward from the winch and gear unit and nevertheless allows the driving beam structure to rotate in a horizontal plane above this unit. This much is adequately shown by the drawings, although it is poorly explained in the description. With regard to other features, the drawings are incomplete. This is mainly so in the component of the ratchet and its pawl,

[8] *Luna di queste funi avogie e lattra isvogie nela ruotta che a e dentti. Ve nel mezo dua girele dove chorono queste. E uni [una] di sotto e di sopra ne dua [?] alttre minori . . . .*

[9] It was copied by Leonardo; see L. Reti, *Tracce dei progetti perduti di Filippo Brunelleschi nel Codice Atlantico*, Florence, 1965, p. 16, Pls. 10 f.

whether this pawl be automatic (as had been usual since Antiquity) or for some reason hand-operated. The illustration is interesting as it gives a clear showing of pivoted gear teeth, called *palei* in the text; they are said to be some 4 inches high and may rotate on iron pins about 1 inch thick. Both drawings and description give much attention to a cover marked d, e, f, g for closing the space containing the rope drum or drums, but it does not become clear why the space should be closed.

The machine seems to be developed for some interesting operation, not necessarily pertinent to the Opera's work, and certainly inapplicable to its heavy work. Lorenzo or others may have made or acquired the drawings and may have offered them to the Florentine Opera and its masters, in which case a more or less polite rejection was probable. The Ghibertis may also have made or preserved these drawings for purposes now unknown.

A REVERSIBLE HOIST

Another and apparently more massive hoist is shown on folios 102^r and 103^v of the *Zibaldone*, Figures 16, 17. It comprises a framework of wood beams, anchored in the ground by wedge blocks. This framework is large enough to let an ox or horse, or several of them, walk around a vertical shaft that is pivoted in the center. Assuming that the parts are shown in at least roughly correct proportion to the actual size of the draft animal, the full height

of this beam structure must be about ten to twelve feet, and the illustrated frame and machine must weigh several tons. It differs from the "secret" hoist in that it will, when properly connected, permit the handling of such loads as the large stone beams of the tie rings used in Filippo's Cupola.

The animal (not shown here) turns the vertical shaft, which then, by reversible gearing, turns a large drum unit about a horizontal axis. This unit in turn, by secondary gearing, rotates a thinner and longer drum or shaft. A rope is very schematically shown as wound on this latter drum or shaft, but obviously a rope could also be attached to the larger drum unit, for example, near the point where this unit drives the thinner shaft, or to the rope drum of largest diameter near the drive gears.

These pages of the *Zibaldone* have no cryptography and no other text. The drawings differ from the preceding ones by their arrangement of wheels and drive means, the absence of a ratchet, and the absence of rollers inside the bearing of a gear. Further differences may be stated more positively by noting the use of a screw mechanism and of three rope drums of different diameter, all strong enough to wind up a long and heavy rope. The Opera documents mention a system of three drums as part of Filippo's hoist and make it clear that the drums were able to wind a very long and heavy rope.[10]

[10] *Cupola*, Doc. 125: *subbio del chanapo, subbio colle ruote*; Doc. 148: *subbio grossiore dicti hedifitii*; Doc. 150: *il subio groso . . . il subio del dificio e di mezo . . . il subio sotile*; also see Doc. 151, etc.

16. Brunelleschi's hoist. Facsimile from *Zibaldone*, BR 228, fo. 102r.

17. Same hoist. Facsimile, *ibid.*, fo. 103v.

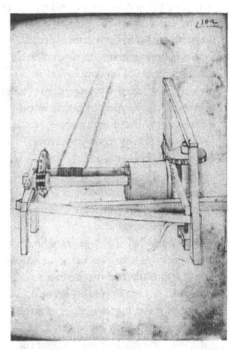

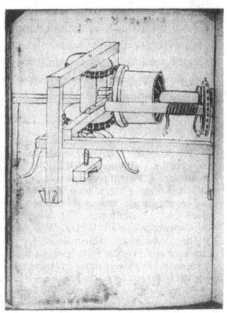

In addition they mention a screw as part of Filippo's hoist.[11] Whatever the details of the screw device and the three drums, the combination is specific, unusual, and characteristic. Figure 17 shows Filippo's machine described by the documents. The conclusion is in no way disturbed by any of the other elements that happen to occur in the *Zibaldone* or the documents.[12] Few inventions of any time have such coherent documentation.

Mariano Taccola was one of those who redrew the machine, in a view rather similar to that of the *Zibaldone*. He also described it:

> The engine with the great wheel . . . is turned by a horse. Let it be known that it is built for three reasons: first, because it turns rapidly; second, because it facilitates lifting, as large weights are raised high; third, because it runs forward and not backward. Fourth, it does not waste time, which is most important in the building of a large structure. If the wheel with lateral teeth is large, the two small pinions fixed on one shaft can easily revolve with rapidity. When the

animal is stopped because the weight has reached the top, then with a rod turn the screw which supports the shaft . . . .[13]

Or in more modern terms, the device can be driven by a horse, as shown. While providing rapid and powerful operation, it has the advantage that it can be reversed without reversal of the drive animal. As Mariano notes on one of the next folios, "The horse always goes forward as he turns the wheel." To unhitch, reverse, and reattach the animal was much more cumbersome and slow and to avoid such waste of time was "most important," according to Mariano. Technically, the screw and gear mechanism is a reversing clutch, one of the basic parts of modern engines.

The reversing feature is mentioned in various copies of Mariano's text and illustration, and important engineering developments appear to follow it. A copy with clear text may be found in a Cinquecento manuscript attributed to Neroni (Figure 35), which has the notation, "These wheels are used one at a time," *lavorano una di queste ruote per volta*. The present tense used in the note may indicate that the device was

---

Also see *Cupola*, Doc. 139: *uno chanapo . . . peso libre 1475*. We can assume that this rope, weighing ca. 1,100 English pounds, was of hemp and that it was about 600 feet long. It then had a diameter of ca. 2.4 inches and could support up to ca. 10 or 15 English tons.

[11] *Cupola*, Doc. 125: *la vite de lo edificio*.

[12] It was necessary to prevent reverse, load-actuated rotation. In Filippo's machine this may have been done by three prongs or *cavigli*, mentioned in *Cupola*, Doc. 125. In the *Zibaldone*, a nonreversible variant of the machine (without screw) appears on fo. 128ʳ.

[13] Cod. Palat. 766, fols. 36ʳ, ᵛ dated 1433, Florence, Biblioteca Nazionale (see Figure 28). Mariano Taccola also shows variants, fols. 37ʳ and 38ʳ. His representation disregards details, such as the arrangement of bearings. When Taccola cites "three reasons" for building the machine and then adds a fourth reason as an afterthought, he seems to imitate Filippo. In Taccola's own formulations such afterthoughts are rare, but they recur in his explicit quotation of Filippo's instructions (Lat. 197, fols. 107ᵛ, 108ᵛ).

still, or again, in use.[14] The reversing gear then reappears in the printed Theaters of Machines, sometimes in forms that appear realistic, although there are fantastic proposals as well.[15]

The Brunelleschian origin of these developments can be demonstrated by comparing the Opera record with the drawings of Ghiberti, Mariano, and others. Even secondary features of the reversible hoist structure can be traced into explicit, Brunelleschian documents. A tiller, to which an animal is yoked, rotates the two pinions fixed on one shaft, called "rotor with wheels."[16] This rotor drives the "wheel with lateral teeth," alternately by the upper or lower pinion, depending on the setting of the screw. The "wheel with lateral teeth" is the free end of a unit constituting a "drum with wheels." The operator can "with a rod, turn the screw which supports the shaft," to reverse the direction of the drums. The reversing elements also include the thin and slow rope shaft, which Mariano does not show, and which is cited in the documents as the "smaller drum."[17]

Some drawings of this machine also show smaller elements mentioned in the Opera documents, for example, the "bearing boxes" for the shafts.[18] Rotatable gear teeth are specified in the documents. These are vaguely shown in Figures 16 and 17 but are not included in Mariano's version. They are shown clearly in the much later Neroni version of the machine. These rotatable teeth must have been a matter of interest, which is also indicated by their use in the "secret" hoist (Figures 14 and 15). Various drawings by Mariano show the use of other antifriction rollers.[19] These rollers may have reduced, although surely not eliminated,

---

[14] Substantially identical copies of Taccola's autograph, Palat. 766, fo. 36ʳ, occur in Palat. 767, p. 173; Cod. S. IV. 5, fo. 84ᵛ, Siena, Bibl. Comunale; Add. 34113, fo. 25ᵛ, B.M. The Neroni codex is S. IV. 6, Siena, fo. 49ʳ. It seems related with San Gallo (see Figure 31) rather than Taccola or Ghiberti.
[15] One of the reversing mechanisms is a rope drive for a car moving back and forth, Francesco di Giorgio, *Trattati*, I, Pl. 105. It also occurs in the *Zibaldone*, in drawings of Antonio San Gallo the Younger, and later in J. Besson, *Théâtre des instrumens mathématiques et mécaniques*, Lyons, 1579, fo. 34, where a 180° drive-gear segment drives a pair of pinions and thereby a pair of winches to reciprocate a cloth pressing carriage. This *artifice nouveau* (?) then recurs in A. Ramelli, *Le diverse et artificiose machine*, Paris, 1588, p. 283, and V. Zonca, *Novo teatro di machine*, Padua, 1607, pp. 34 f. A closer variant of our Figure 28 is in G. Branca, *Le machine*, Rome, 1629, part I, fols. 20ᵛ, 21ʳ, and 30ᵛ, 31ʳ—see our Figure 36.
[16] *Cupola*, Doc. 125: *lo timone de' buoi*, tiller; *le ruote a charuchole*, rotor with wheels, rotating in two *bilichi*. The gears of this unit had roller teeth, *palei*. Doc. 125 also lists, and the drawings show, the additional parts mentioned in this paragraph.

[17] The bearings for this shaft needed specially smooth surfaces because this shaft raised the heaviest loads. A pair of walnut sleeves, *girelle di nocie*, was provided, probably for these bearings. For the "smaller drum," see *Cupola*, Docs. 150, 151.
[18] According to *Cupola*, Doc. 125, four such boxes were made from twelve feet of oak wood.
[19] The unpublished manuscript in Siena, Biblioteca comunale, Cod. S. IV. 6, fo. 49ʳ, attributed to B. Neroni, which calls the *palei* by the name *rocchetti*, shows a total of 102 of them (24 + 24 + 24 + 30), while *Cupola*, Doc. 125, speaks of 91. Taccola in Lat. 197, fo. 1ʳ, shows a roller frame for facilitating the handling of tree trunks or columns or cannon tubes.

the large friction losses of the cog-wheel drive.

With regard to some small parts, the secret hoist is described in terms which, at first glance, appear closer to the Opera record than Figure 17. These parts are called "rope-guiding rollers." They are not shown in drawings of the Brunelleschian hoist but may be found in the official, Brunelleschian record in the Opera (Doc. 125). The explanation probably is that Filippo's rope-guiding rollers, in contrast to those of the secret hoist, belonged to the elevated crane structure. We must turn to this structure.

ELEVATED CRANES

Cathedral builders of the High Gothic age in northern countries had generally used a man-powered hoisting wheel, which they placed in the roof of the cathedral. A typical sequence of operations consisted of constructing the foundations, the buttressed walls and piers of stone, and the roof structures of wood resting on the top edges of walls and piers; assembling hoist winches in the roof; building stone arches with the aid of the winches; placing vault masonry on supports suspended from the arches; and then placing the glasswork between the piers of the outer enclosure.[20] It is conceivable, although not very probable, that a similarly sophisticated sequence was used by Ghini when, in the 1360s, he built the nave of Santa Maria del Fiore.

[20] Fitchen, *The Construction*, p. 139.

For one thing, the documents only indicate that he used a centering, and the terminology is not always distinct. In any event, a sequence as used in the tall, narrow cathedrals of the North could not be used in the Florentine Cupola, as no one would construct a roof or arches across the wide span of 140 feet. Gothic construction prior to Filippo had been of breathtaking height, but never had it covered a span of such breadth. The Gothic predecessors of Filippo in Florence may have used machines that were more advanced than mere winches (for example, when building the Tambour), but nothing is now known about them or their elevated crane structures.

The biographers do not say how Brunelleschi designed such elevated structures, but the *Zibaldone* has pertinent suggestions, shortly after the pages showing the ground installation. The first of these suggestions, in the *Zibaldone* sequence, is a rotary crane similar to a merry-go-round, Figure 18. At its top there is a grooved, horizontal beam structure wherein a screw-engaging block can slide. The block is shifted along the groove by a horizontal screw. There is also a vertical screw passing through the block, and this latter screw can support three hangers, shown separately on the same page. The device was also shown by Leonardo.[21] Evidently it is a load-positioning machine, standing on a roller platform.

[21] *Cod. Atlanticus*, Ambrosiana, Milan, ed. G. Piumati, Milan, 1894–1904, fols. 295r-b, 295v-b; Reti, *Tracce*, p. 23 and Reti's Fig. 24.

18. Elevated rotary crane. Facsimile from *Zibaldone*, BR 228, fo. 104ʳ.

19. Rotary crane for Lantern top. Facsimile, *ibid.*, fo. 105ʳ.

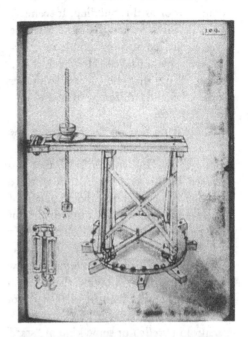

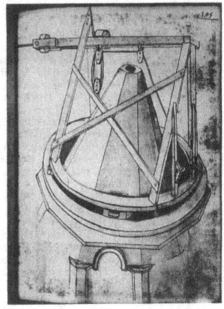

A drawing on the next folio, Figure 19, shows a very similar unit. In this drawing the machine evidently stands on Brunelleschi's Lantern, around the conical roof of this building. It seems possible that one and the same machine was assembled and reassembled on successively higher levels. However, it is also possible that the successive cranes were of slightly different form and of different size.

There is no text on these pages of the *Zibaldone*, while a corresponding drawing by Giuliano da San Gallo, also copied by his nephew Antonio, has a note of some interest.[22] Some details of the note are not free of difficulty,[23] but so much is certain that Giuliano speaks of a machine for the Florentine *Duomo*.

The Opera record also contains a few pertinent entries, showing that a *stella della cupola* was built in 1420; it is not clear whether the word *stella* meant a scaffold (*'stello*) or some kind of "star." There were wheels to "draw" this

structure (*tirare la stella*) or perhaps merely to draw masonry material onto it. This scaffold may have supported the rotary crane, Figure 18, at one time or other, by the adjustment screws shown in the San Gallo drawing.[24] In this case we may visualize the device of Figure 18 as standing and operating on top of the scaffold of Figure 8. However, the scaffold may also have had a mere rope wheel or pulley suspended from it or supported on it; the documents also speak of a "drum or wheel on the *castello*." The nature and sequence of these devices will be considered again when we study the written record in its chronological sequence. The *Zibaldone* shows a selection of elevated cranes that could have been used, but fails to show whether or at what times they were actually used.

For the Lantern construction we also have a very realistic drawing of unknown origin, showing machinery quite different from that of the *Zibaldone* pictures. We show this Lantern crane in Figure 27. It may be an authentic record, traceable to the Lantern-building operations performed during the second half of the Quattrocento. It has in its upper right corner a note in a hand of that time, indicating that

This last scaffold reaches up to [the foot of] the *palla* [or gilt brass orb on the Lantern]. When the *palla* shall be raised [as was ultimately done by Verrochio and young Leonardo]

[22] About Antonio's use of Giuliano's drawings, see the brief notes of G. Giovannoni, *Antonio da San Gallo il Giovane*, Rome, 1959, pp. 28 ff., 70 f. The drawing by Antonio San Gallo the Younger (Uffizi, A 1664ᵛ) has the note, *Chome si muro la lanterna della cupola di Sta. Maria del Fiore*. On his next page, Uffizi A 1665ʳ, is a reference to the *libretto di Giuliano*. A somewhat similar reading is in Stegmann and Geymuller, *Die Architektur*, I, p. 48n, based on Giuliano's own *taccuino*; see note 23.

[23] R. Falb, *Il taccuino senese di Giuliano San Gallo*, Siena, 1902, fo. 12 of Cod. S. IV. 8, reads *Chome si mu(ove) la champana de la chupola di S. Liperata*. The reading, *murò*, of Antonio San Gallo and Stegmann, as given in the preceding note, is more plausible. The word *champana* is probably vernacular for Lantern.

[24] *Cupola*, Docs. 122, 135 f., 171, 275 f.; also see Doc. 304.

another scaffold will be needed, about 23 feet high.[25]

The note indicates that this is neither the first nor the last of the scaffolds used in the Lantern area.

In a similar way the scaffolds probably underwent periodic change during the earlier phase of the work, when the Cupola itself was raised. There must have been scaffolds even before Brunelleschi started his Cupola construction. It appears that he added to them as soon as he began this work. Some three years later reference was again made to scaffold construction work to be done, *pro fiendo castello*.[26] The hoisting machine, Figure 17, seems to have operated for more than a decade without major change, but the elevated scaffolds and no doubt the crane elements thereon were subject to relatively frequent change.

In addition, variants of the machines were proposed by workers and advisers, and it appears that their forms may be found in the notebooks.[27] In some cases it is beyond proof that the device was actually used, or could be used to advantage, even if a statement suggesting its use occurs in a book such as the *Zibaldone*. In Figure 24, we see a screw mechanism and read that "they turned the Cupola scaffold with it," *chon questo si vo[l]gieva el chastello della cupola*. More plausibly, it may mean that by turning the device, they adjusted loads while operating on the Cupola scaffold. However, it is conceivable that a scaffold of the type as shown in Figure 18 was turned, slowly, by screw action. Several notebooks show devices for this purpose.

Still other drawings add a man-operated winch to an elevated crane device instead of relying on the floor-mounted winch.[28] Again there is no positive showing in the notebooks or official documents that such a modification was actually used. The impression arises that no one but Filippo and his direct helpers knew the complete record and plan with regard to these matters. The Opera tried to know clearly what was planned and proposed with regard to the masonry, and it had considerable difficulty in understanding this procedure. It hardly tried to record details of the machinery. These details are found only in the surviving drawings. They were as unintelligible to Vasari as they were to the Opera.[29]

[25] *Questo ultimo ponte viene appunto alla palla sicche . . . bisogna alzare uno altro ponte alto br. 12 in circa*. The phrase represented here by an ellipsis is not clearly legible but seems to say *a volere condur la palla*. "This last scaffold reaches to the orb, so that in order to mount the orb it will be necessary to raise another scaffold, ca. 23 feet high." We were unable thus far to establish the watermark of this drawing, which could confirm or modify the assumption that we have here a Quattrocento record. Also see *Cupola*, p. 201.
[26] *Cupola*, Doc. 134; also see Doc. 126, of the same year, which reads, *da fare el chastello*.
[27] For example, Cod. S. IV. 6, fo. 32ʳ reproduced in Scaglia, *J.S.A.H.*, Fig. 6.

[28] *Trattati*, I, p. 195 and Pl. 95; British Museum, Harley 3281, fo. 12ʳ; Turin, Cod. NL 383, fo. 19ʳ; Siena, Bibl. Comunale, Cod. S. IV. 1, fo. 126ᵛ, S. IV. 5, fo. 17ʳ; Florence, Palat. 767, p. 201. A close variant is NL 383, 38ᵛ and Palat. 767, p. 190 (see Scaglia, *J.S.A.H.*, Figs. 1, 2, and *infra*, Figure 30).
[29] *Vasari*, pp. 203, 220, 222. The later

During early stages of the Cupola construction, ca. 1421 to 1430, devices were needed whereby workers could transfer heavy loads, such as stone blocks for the tie rings, to the eight corners and sides of the Cupola. It was necessary to use a central mechanism for hoisting these materials. When these heavy materials arrived in upper, central location, it was impossible to transport them by hand over the Cupola scaffold, even if the entire work force was employed. Yet it was necessary to place these blocks in accurately defined positions, interlocking with other blocks, with micrometric accuracy. It is technically obvious that this was the purpose of devices additionally shown in the *Zibaldone*, for which Figure 20 is typical. In later stages similar functions could also be performed by the rotary cranes, as already indicated.

The load positioner has two horizontal, screw-actuated slideways, apparently destined, respectively, to manipulate the live load and a counterweight. The two slideways are mounted on a vertical rod that can turn in or on a larger vertical post, erected somewhere on the Cupola walls, as indicated in Figure 26. A "tiller to draw the device," *timone*

---

statements by Manni, *De florentinis inventis*, pp. 79 f., indicate a recognition of the fact that positioning of the heavy building elements was necessary: *ingentibus in lapidibus marmoribusque a solo efferendis unumque super alterum statuendis*. However, Manni seemed unaware of the machines for such positioning.

che tira il dificio, turns the slideways into the proper orientation. Operators then turn the horizontal screws to shift the load into position above its ultimate place and correspondingly to adjust the counterweight. They then manipulate the threaded nut engaging the vertical screw to lower the load into place; a horizontal pin prevents this screw from rotating. Prior to this stage, the load had to be transferred to the lower end of the vertical screw, after it had been hoisted by the ox-driven machine. All this is fairly obvious from the drawing alone; its text is of no assistance at all. It speaks, with prolixity, only of the vertical shaft, which

> . . . is turned by the tiller on the [part marked] +. [It is] on [this part, and it extends] from + upwards. Thus [being turned by the tiller], it begins to turn where these rollers are, on this circle, in the middle of the device . . . .

This apparatus, like most of the Brunelleschian machines, is also shown in Leonardo's notebook (Figure 34), and there are other versions (Figure 30, probably traceable to Francesco di Giorgio). Contributors to the Leonardo literature have been inclined to use exclamatory statements and have been unduly impressed by the progress that they believed to find in this device, which they promptly credited to their hero.[30] They were unaware of the technical developments attributable to the early Quattrocento and to still earlier times reaching back to Antiquity.

[30] Cod. Atlanticus, fo. 349r-a; see W. Treue, *Kulturgeschichte der Schraube*, Munich, 1955, frontispiece and p. 81.

20. Load positioner, Facsimile from *Zibaldone*, BR 228, fo. 106ʳ.

21. Load positioner and crane. Facsimile, *ibid.*, fo. 107ᵛ.

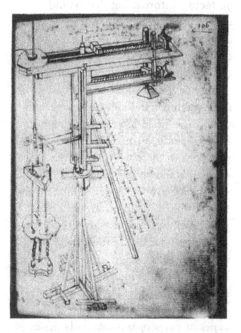

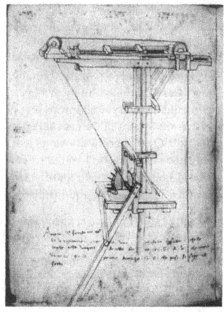

The real question is whether Filippo invented or improved such load positioners; it is clear that Leonardo only noted them as he found them. We have no factual information that would enable us to answer the question about Filippo's contribution to technical progress connected with the design of the machine. Only so much seems clear, that he devised a sophisticated system of floor-mounted and elevated machines using these precision mechanisms.

The *Zibaldone* has still further entries, pertinent to these developments. One of them is shown in Figure 21. Obviously it shows a device similar to the first peripheral load positioner and is additionally provided with a rope-actuating winch. The book also shows a machine having the slideway and its double-screw device, mounted on a kind of carriage.[31] These double-screw devices are shown in so many different types of carriers and manipulators as to suggest that the screw device was, or became, a unit of some little notoriety.

In the documents of the Opera we may find an old name for such a device. Reference is made to *mozetti*: literally, screw vises.[32] Although the word by itself could be interpreted in various ways, it seems most fitted (as applied to any of the Quattrocento drawing elements known at present) to the screw-actuated nut elements of Figures 18 to 21, which indeed act in the manner of

vises although they are used as sliders of a load-carrying machine.[33]

OTHER DEVICES

Other illustrations in the *Zibaldone* show an assortment of auxiliary tools and other machines. Figure 24 shows a pair of screw-acting mechanisms, probably used as load hangers. Figure 25 once more shows the triple hanger of Figure 18, this time attached to the vertical screw of a load positioner. It also shows a set of stone-engaging keys or hangers, of very ancient type. Specimens of some of the hangers are preserved in the Opera.[34] They became very famous, as they are specially mentioned by Vasari, but we doubt that their interaction with the various hoisting and positioning machines was clear to the writer of text notes or remarks in the *Zibaldone* pages devoted to these devices; the notes and remarks do not show it. We also doubt that this system became clear to many observers, except a master craftsman such as Leonardo. He illustrated the positioners next to the hoist and clearly showed them as interrelated machines (Figure 33); he seems to be the only one to show them so, in spite of the dozens of variants

---

[31] Scaglia, *J.S.A.H.*, Fig. 4. The drawing also appears in Palat. 767, p. 207; *Trattati*, p. 194, Pl. 94; Harley 3281, fo. 13ᵛ; NL 383, fol. 21ʳ; S. IV. 1, fo. 129ʳ. Also see B. Gille, *Les ingénieurs de la renaissance*, Paris, 1964, p. 99.

[32] *Cupola*, Docs. 125, 133, 140 f.

[33] It is conceivable that *mozetti* cooperated with Filippo's *vite*, but it would not be clear in that case why several casings were provided for them. Still other interpretations and suggestions, in *Osiris*, IX, pp. 509 ff.; *Marsyas*, X, pp. 53 f., 56, are now obsolete.

[34] *Il progetto*, Figs. 10 ff.; Reti, *Tracce*, Figs. 4, 7; also see *Cod. Atlanticus*, fols. 10ᵛ-b, 339ᵛ-a, 389ᵛ-b.

22. Theater machine (motor and lamp unit). Facsimile from BR 228, fo. 115^r.

23. Same machine (frame). *Ibid.*, fo. 115^v.

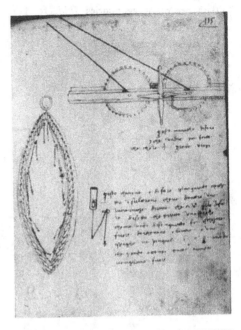

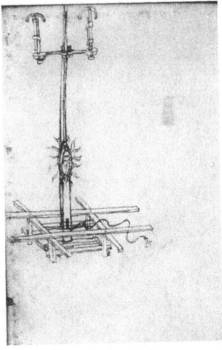

24. Crane accessories. Facsimile from *Zibaldone*, BR 228 fo. 117ʳ.

25. More crane accessories. *Ibid.*, fo. 119ʳ.

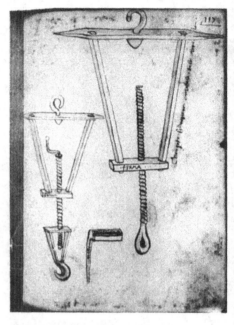

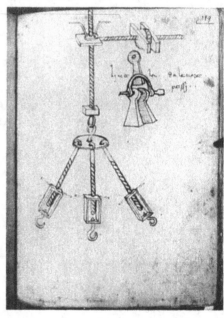

produced by Mariano Taccola, the San Gallos, Francesco di Giorgio, Buonaccorso Ghiberti, and others. Generally speaking, the *Zibaldone* and other notebooks fail to show the machinery of Filippo in its entirety. Even Leonardo presents only an unfinished sketch of the complete system.

The *Zibaldone* contains a few drawings that, in the absence of meaningful description, cannot be interpreted with assurance. Among them is a device that may have been used for measuring altitude angles (fo. 117^v) and a device shown with balanced buckets, which may be intended as an adjustable support for masonry (fo. 119^v). These devices may be worth further study. At the present point we mention them only to give a reasonably complete account of the machinery for the cathedral work as recorded in the *Zibaldone*.

Another pair of *Zibaldone* drawings, Figure 22, 23, has no parallel in other notebooks now known. A crank turns a lantern gear and thereby actuates a pair of larger gears, which have shafts winding up ropes that lead to an almond-shaped frame equipped with small tubes. A detail view and note make it clear that the tubes contain candleholders. Each tube has a cord whereby its candleholder can be pulled into the tube against the compression of a spring. In order to expose several of the burning candles to sight, the operator turns the wheels, thereby causing the cords to relax the springs. The device seems to be part of a Brunelleschian festival apparatus, built for "Paradise" representations in San Felice in Piazza. Vasari describes the general arrangement of the machine. The compression springs, an element extremely rare at the time, may be connected with other Brunelleschian developments.[35] However, we must turn to a more systematic review of the writings connected with these assorted drawings.

[35] *Vasari*, pp. 229–232. About springs in Brunelleschi's time: Prager, "Brunelleschi's Clock?" pp. 268 f. About another system of lamps, going on and off successively, a "secret" description may be found in Giovanni Fontana, *Liber instrumentorum bellicorum*, Munich, Bayerische Staatsbibliothek, cod. icon. 242, fo. 20^r: *Nocturnus ignis qui videntibus mira ostendit. Gradetim ponantur ceri usque ad lampadem qui sucesive accenduntur unum post alterum sicud scivisti. Et serpitarium cereum fit deperse in capsa sua quod ultimo accenditur. Deinde eum illud agere coneris quod de umbris in claro et luminibus in obscuro facere me vidisti.*

HISTORY OF BRUNELLESCHI'S MACHINES

5 THE ORIGINS

Manetti describes mechanical studies of Filippo, along with his Roman discoveries. No dates are cited, but the impression is created that all of these studies and discoveries occurred soon after the Battistero contest. Vasari writes, more plausibly, that some time elapsed between the basic studies and the development of new and complex machinery.[1]

The biographers indicate that Filippo studied machinery by inspecting actual structures. Others have suggested an influence of ancient, Vitruvian teachings,[2] but no evidence has been adduced for such an influence in the field of machinery. Poggio discovered a well-preserved and supposedly intelligible manuscript of *De architectura* in 1417; it was copied in later years. The discovery became very sensational,[3] but even if it made the ancient mechanical teachings understandable, which may be doubted, it came entirely too late to influence Filippo's planning of mechanical devices. No doubt Filippo may have known of other Vitruvian manuscripts owned by Florentine humanists, but their terms, confused by medieval copyists, were substantially unintelligible.[4] Some friend could have translated, for Filippo's use, the more accessible *Epitoma* of Faventinus, but there is no evidence that such a translation was made. It rather appears that Filippo, as a mechanic, was guided almost exclusively by inspection and consideration of then-existing machinery.

BUILDING THE MACHINES

In 1417, models of machines as well as models of the Cupola were submitted to the Opera. It seems probable that Filippo contributed a model or several models, or designs for them. No doubt Filippo knew the background of prior experience with machines used by builders, first through his father's influence and later through his own contacts as consultant to the Opera. Often during the Trecento, when building operation entered into a new phase, artisans developed new hoists, including man-operated or animal-operated winches to hoist materials to the height of the Campanile, or later to

---

[1] *Vasari*, pp. 202 f. His account of the machinery, pp. 203, 220, while almost as meager as *Manetti's* account, p. 19, adds a remark about devices called "hinge hooks." Those persons who saw the machines and described them to Vasari may have been dimly aware of the load positioners hinging about vertical pivots.

[2] *Fabriczy*, p. 40; F. Pellati, "Vitruvio e il Brunelleschi," *Rinascita*, 1938, pp. 343 ff.; *Sanpaolesi*, pp. 28, 32.

[3] E. Walser, *Poggius Florentinus. Leben und Werke*, Leipzig, 1914, pp. 53 f.

[4] A relatively well written, twelfth-century codex of Vitruvius, Urb. Lat. 293, Biblioteca Vaticana, offers the following designations of machines: (80ᵛ) . . . *alterum spirabile quod apud eos [grecos] pe-umaticon appellatur. tertium . . . barvison . . .* (81ʳ) . . . *uallistae . . .* (82ʳ) . . . *sucularum venationibus . . .* etc. Even the correct terms (*pneumaticon, barylcon, ballistae, sitularum*) were difficult; the terms actually used by the copyist were hopelessly confused. The text is from X, i, 1.

the Tambour.[5] Some such machines probably were in use when Filippo was a child.[6] Now he began to improve them.

In the summer of 1418, just before Giovanni d'Ambrogio was dismissed, the construction of scaffolding for the Cupola was initiated. Document 237 of 6 July 1418 authorizes

| 32 logs of large tree trunks to make boards for bridges (pontes) for vaulting the Cupola.

Although details are unknown, perhaps these boards were to provide a brick-supporting centering, not a system of mason's walkways. Certainly the thirty-two tree trunks were only a small beginning toward an armature for a centering. No such construction was built, as Giovanni was dismissed.

The next step toward construction of the Cupola came at the end of April 1420, before Brunelleschi's specification was formally accepted. The Opera then decided that

| arrangements shall be made to prepare . . . 100 fir trees, hauled in bundles of 6, for making bridges (pontes), each tree 21 feet long.

These were in addition to materials

and "a wheel for pulling the materials" (Doc. 238). The next records speak of centerings and iron parts:

| 28 June 1420. [The Opera authorized payment of] 4 lire, 8 [soldi] for manufacture of eight centerings of fir by Mariano Benedicti, wood worker (Doc. 170)

and

| 9 July 1420. Nanni di Fruoxino, blacksmith, shall have [payment for] 132 [Florentine] pounds of an iron chain for the stella [scaffold?] of the great Cupola, and 280 pounds of iron plates for iron-trim of the centerings for said Cupola (Doc. 171).

The iron chain may have been used to hold some parts of the scaffold together (Figure 8). The early book entries of the Opera relating to Brunelleschian work then continue with a notation about a modest feast, celebrated "on the morning when construction (murare) of the Cupola began" (Doc. 239, 7 August 1420). The masters consumed some wine, bread, and melon. This record, dating from April to August, is reminiscent of transactions made ten years earlier when some bricks were delivered, for ten soldi, at the start of the Tambour work. After the feast, work continued:

| 8 August 1420. 1 soldo for two wheels for pulling [on?] the stella of the Cupola, taken by Pippo di Ser Brunellescho. [Paid] 4 September 1420 (Doc. 122).

It is clear that Filippo had completed the design of the elevated woodwork, that at least part of it was in place, and that, with its aid, elevated masonry work had begun.

---

[5] Costruzione, Doc. 70 at pp. 82, 85, 90, 111, 116, regarding hoists of Maestruzzo, Bartolo di San Gallo and Jacopo di Vanni; ibid., Doc. 72 at pp. 121 ff., regarding a hoist of Ghini, expected to "save money each day" by eliminating "work of tread-wheel operators," per lo quale l'Opera risparmierà ogni dì lire 3 soldi 3 . . . vi bisognerà 7 manovali meno che nelle due ruote. Also see for example, Holt, Documentary History, Figs. 12, 13.

[6] Especially a machine built in 1380 by Filippo Justi, with horse drive (Costruzione, Doc. 314). It probably scooped water from the foundation area (Figure 4).

Three weeks later Filippo began personally to make payments to individual artisans for the various parts of the ox hoist, which he furnished to the Opera as a contractor. Since he later obtained itemized reimbursement for these payments, this phase of the work is unusually well documented. The record shows that he used a great many artisans, evidently under a fully developed, preexisting master plan, and that he must have worked furiously for several months to keep these artisans in line and thereby to produce his famous, smoothly operating machine. This machine-building took place about the same time when the arcades of his *Innocenti* porch began to revolutionize Florentine city planning. The payments for the machine are summarized in a single payment voucher (see Document Section). Part of it may be translated as follows:

26 August to Montino di Bruogio for hauling an elm tree for the rope drum: 3 *lire*. 30 August to Testa, customs duty for two wheels: 20 *soldi* . . .

and so on, for the initial deliveries of materials, including "oak wood to make bearing boxes . . . walnut sleeves for the rope shaft . . . chestnut posts to support the machine . . ." and notably, customs duty for "the screw." Also furnished or constructed was a "piece of chain, . . . harness for the oxen, . . . further wood work . . . and rope."

Finally we come to a paragraph relating to major work elements and their assembly:

10 March [1420/21] to Fruosino d'Andrea, cooper, for 14 hoisting tubs . . . 40 *lire*, 12 *soldi*. To Mattio, blacksmith, for 2 bearings for the pulley wheels . . . and to Maxo di Chiricho, cartwright, for various woodwork that belongs to the machine: 11 florins, 56 *lire*. To master Antonio Stoppa for manufacture of the screw (*vite*) for the machine: 5 *lire*, 2 *soldi*.[7] To Antonio di Tuccio, wood turner, for manufacture of 91 roller teeth (*palei*) at 4 soldi each, and for 16 screw-vise blocks (*mozetti*) at 8 soldi each, and 6 lire for wood: 30 *lire*, 12 *soldi*. To masters Piero and Antonio de' Bianchi, who respectively spent 67⅓ . . . and 67 days . . . in building the machine: 151 *lire*, 1 *soldo*, 8 *denarii*. And to Giovan di Fruosino, blacksmith, for 1022 pounds of various iron trim . . . and for making 2 bearings for the upright wheel . . . : 229 *lire*, 4 *soldi*, 8 *denari*. Total: 11 florins, 584 *lire*, 12 *soldi*, 5 *denari*. (Doc. 125)

The principal items of this list (more fully shown in the document section of this book) are clearly reflected by the Ghibertian drawing, Figure 17. It shows the tiller, or *timone*, turning the vertical shaft with the pair of drive gears, *ruota a carucole*. It shows the horizontal gear and winch unit, *subbio colle ruote*, the left end of which engaged one of the drive gears, while the right end had a lantern wheel driving the second,

[7] Unfortunately the documents do not indicate the material of the *vite*. The device was made by *Maestro* Antonio Stoppa, otherwise unknown. We incline to think it was made of wood. If made of metal, it would weigh hundreds of pounds, and would cost considerably more than the 5 lire, 2 soldi actually paid. About related developments of the time, see Scaglia, *J.S.A.H.*, Figs. 7–15, 18, and *ibid.*, p. 91, note 2.

thinner, and more powerful shaft, the *elmo*, *stile del canapo*, or *subbio minore*. This is a unique and unmistakable combination, shown with clarity in Filippo's expense bill and the Ghibertian, Leonardian, and other drawings. For good measure the Ghibertian and other drawings and the Brunelleschian bill coincide in the inclusion of another unmistakable feature, the screw or *vite*.

The drawings make it clear what function each listed element served, while the documents show the sequence of machine constructing operations in which the illustrated elements were united. Preparatory work, relating to the elevated machinery, began in July 1420, as we have seen. From the end of August 1420 until October of that year, Filippo received parts for the ox hoist from artisans outside the Opera, some of them outside Florence. Their work, no doubt, was done on the basis of specifications given to each supplier. Thereafter and until March 1420–1421, Filippo and his helpers in the Opera assembled and supplemented these parts to produce his hoisting machine. The machine then stood in the center of the octagon for the next twelve years, if not longer. It was as heavy and rigid as a house and was more versatile than machines produced until then. Meanwhile the Opera, without letting out contracts as Filippo had done, had built the scaffold and centering system needed in accordance with Filippo's building method. It is clear from the document of 8 August 1420 that Filippo was in close contact with that operation, although he did not furnish it as contractor. As indicated by his

reference to screw vises, *mozetti*, he probably furnished the load positioners that constituted the principal new and specialized part of the elevated structure. The entire system (Figure 26) was predicated on the use of his method of vaulting without armature, as it occupied the space normally taken by an armature.

WORK OF THE MACHINES

In the early spring of 1421, when the machines were complete, Filippo submitted his claim for two payments: a prize for the entire system, and compensation for his expenditures for the ox hoist. On 10 June the Opera promised compensation for the expenditures, on 18 July it awarded a prize of 100 florins, and on 20 August—about one year after the contract work began—it paid the compensation, which came to almost 200 florins.

Meanwhile the Cupola administrators made contracts with operators of the ox hoist, beginning in March 1421, when they

> deliberated that Bartolinus Bartolomei Cagnani, who works with his oxen on the device of Filippo di Ser Brunelleschi to pull stones and other things onto the major Cupola, etc., shall have 31 *soldi* [for unspecified time or piece work] when he works with one ox and 50 *soldi* when he works with two (Doc. 146).

Later contracts stipulated similar amounts per day. Thereafter, 1421–1423, the Opera stipulated certain amounts for each load hoisted. Still later it returned to daily wages (Docs. 146–157, 249). We may safely estimate that the oxen raised an average of some fifty loads, ranging from a few hundred pounds to several tons, during a typical workday.[8] During the twelve years of actual use in building the Cupola, the machine raised several tens of thousands of tons of masonry material, not counting its reported use for raising and lowering the workers.[9]

The prize award documents call the machinery (no doubt including the ox hoist and perhaps also the elevated structures) a "new device or hoist constructed for pulling and moving stones, blocks, and other requirements, . . . newly invented by him (Filippo), whereof the Opera has more useful

return than from that previously used." (Doc. 123 f.)[10] No attempt was made to pin the prize or the invention to any particular part or to compare it with a specific machine of the past. For such comparisons we can, however, use a variety of other documents and drawings, as we have done. The documentation shows that the machinery, built about the end of 1420, was a spectacular success.

It operated well over ten years. The ox hoist still was to be seen and was an object of Florentine pride when the San Gallos made drawings of it, ca. 1470 or later. Very few repairs were needed, and there was no complaint. We can be sure that the documents would record any difficulties encountered.

Reference was sometimes made in the documents to a "major hoist," possibly meaning a "larger hoist" and implying a "smaller one." If we knew only the ox hoist, we would conclude that probably others were built. However, there were elevated hoists. Some of these, standing on top of the Cupola walls, must have been among the minor hoists inferred by the expression "major hoist." We will see further evidence to this effect.

Near the end of the first year, we read of a repair, an isolated case:

---

[8] According to the pertinent documents (*Cupola*, Docs. 146, 148, 150, 151, 153, and 155), the payments to the ox drivers were as follows:

March	1421	31 to 51 *soldi* [per day?]
Nov.	1421	6 to 10 *den.* per load
Aug.	1422	6 to 16 *den.* per load
March	1423	7 to 14 *den.* per load
May	1425	42 *soldi* per day
Dec.	1425	40 *soldi* per day

Assuming fifty hoisting operations as a daily average, the reported figures lead to the following average payments: 41 *soldi* per day under Docs. 146, 153, 155, and 40.5 *soldi* per day under Docs. 148, 150, and 151.
[9] The weight of the Cupola is not officially reported. We estimate it, without the Lantern but with the key ring, at ca. 25,000 English tons. During the twelve years, 1421–1432, Filippo hoisted this approximate amount of material to an average height of ca. 250 feet above ground. We estimate an average daily load of ca. 10 tons and a maximum daily load of ca. 25 tons. With the use of such lifting capacity, a pair of tie rings, consisting of stones and iron clamps fully prepared on the ground, could be installed in a few weeks. Parsons, *Engineers*, p. 596, assumes a weight of 70 million pounds.

[10] *Cupola*, Doc. 123: . . . *hedifitium novum construi fecisse . . . pro trahendo et conducendo super muris cupole maioris lapides, macignos et alia opportuna et propterea multas expensas fecisse . . . de quibus et etiam de eius labore et industria nondum . . . satisfactum fuit . . . .* Doc. 124: *per lo suo ingiengno et sua faticha durata, dello edificio per lui nuovamente trovato per tirare, del quale a l'Opera ne torna più utile che di quello che prima s'aveva . . . .*

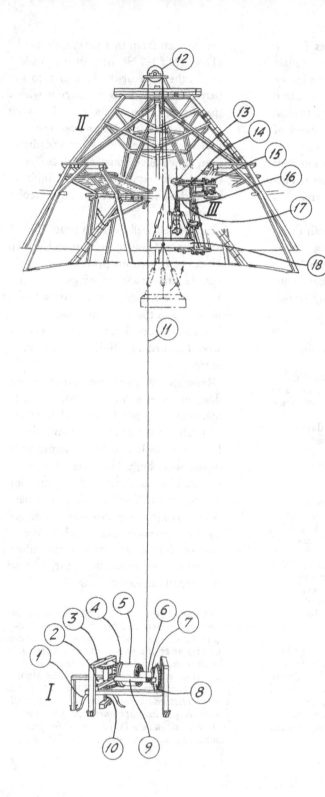

26. Brunelleschi's hoisting system. Drawn by G. Rich, 1969. *1.* Tiller, *timone. 2,3.* Drive gears, *ruote a charuchole. 4–7.* Driven gear with winch unit and secondary driving gear, *subbio colle ruote. 8.* Secondary driven gear (*ruota ritta?*) *9.* Secondary winch, *stile del canapo. 10.* Clutch screw, *vite. 11.* Rope, *canapo. 12.* Pulley, *ruota. 13.* Positioning "vise," *mozetto. 14.* Vertical screw passing through the "vise." *15.* Horizontal slideway for the "vise." *16.* Rotatable post for the slideway. *17.* Antirotation guide for the vertical screw. *18.* Stone block that has been hoisted (broken lines) and is being raised (arrow) for attachment to the positioner.

Bartolomeo di Stefano, kettle maker, shall have . . . [payment] for 267 pounds of bronze worked into eight bearings for the cases (*bilichi per le chasse*) of the pulling hoist. 4 October 1421 (Doc. 130).[11]

On 14 January 1422, in anticipation of the second year of operation, announcement was made that whoever wished to work as ox driver when the current contract expired should register his name. A new driver was hired on 4 August (Docs. 149, 150).

Then came a few further repairs or replacements of small parts. A payment was made for a "wheel or pulley" that friar Antonio di Bartolino had "lost." New chains and iron tools were acquired from "the blacksmith." Also acquired were "three pieces of oak for making three cases to hold *mozetti.*" The latter cases were for a new "device that Filippo constructs" (Docs. 131–133, June to November 1422). From these documents and the drawings we may conclude that new load positioners began to take shape.

The year 1423 began with a new ox

[11] One of these bearings, weighing 34 Florentine pounds, consisted of ca. 145 cubic centimeters of bronze. It could be a hollow casting about 8 cm long and about 7 cm outer diameter. Such a bearing may have been sufficient for the rope drum of an auxiliary hoist on a load positioner. The document speaks of *bilichi per le chasse*, probably meaning "bearings for the load positioner," *bilichi per le casse dei mozetti.* (Contrary to views expressed in *Osiris*, IX, p. 515, and by *Sanpaolesi*, pp. 106 ff., and as pointed out by Scaglia in *Marsyas*, X, p. 54 n, *bilichi* in these documents clearly means bearings, although in other contexts it can mean different things.)

driver's contract (Doc. 151) and continued with further evidence of changes in the elevated structure:

Two wood beams for the scaffold to be built on the great Cupola, each 36 feet [long]: 29 *lire*, 10 *soldi*. 15 April 1423 (Doc. 134).

On the same day (Doc. 126, part I), Filippo received a new prize, this time only 10 florins "for his efforts and invention of the device for the scaffold, for pulling the weights up onto the walls." A few months later, 27 August 1423, he received another 100 florins for several structural and mechanical achievements, including "machines to be built" as well as structures already completed. The popular account of these events, recorded by Vasari, does not go far beyond the fact or assertion that by means of various machines "one ox could raise what six pairs could scarcely have raised before." Vasari also knows, and the documents confirm, that others were paid for "efforts and invention applied to a device for the *chastello* on the walls" (Doc. 126, part II).[12]

Then in 1423 we read about a new rope,[13] and a new wheel (Docs. 135, 136). Both were quite possibly for the elevated load positioners, equipped with winches, which are shown in Buonaccorso's drawing, Figure 21. The documents now speak of an elevated area where the main rope "turns to the ox hoist" (Doc. 137), perhaps indicating that the hoist rope now was periodically pulled, by machinery, toward the load positioners.[14]

During the fourth and later years the Opera recorded only minor matters. A German, called Gerhard, "made a mode[l] intended to hoist like the other device," and when he had left town, his innkeeper presented a bill (Docs. 127, 128). Masetto, the master wheelwrigh[t] outside the gate of San Frediano, who had worked for Filippo before, provided "40 teeth of ash wood," apparently for a cogwheel (Doc. 137, 2[0] December 1424). When the workers reached the level of 58 feet above th[e] Tambour, at which point methods an[d] plans were to be reconsidered, the program was continued with only slight modifications (not affecting the hoisting system).[15] Only the usual ne[w]

[12] Also mentioned by *Manetti*, p. 51, and *Vasari*, p. 218.

[13] Made by one Matteo, who is called *schodellaio*. *Cordellaio* may be meant.

[14] In the Latin text of *Cupola*, Doc. 126, the wording of the prize award refers t[o] both scaffolds and hoists in the plural: *pr[o] eius labore et inventione castelleorum pr[o] collis pro cupola magnia*. It seems possib[le] that the tie rings of the middle layer we[re] constructed with the aid of a load positioner, standing on a scaffold between th[e] shells. Such scaffolds were also needed f[or] centerings of the *volticciuole*. The positioner may have been transported from on[e] side to the next on consecutive days. Th[e] drawings in the *Zibaldone* (Figure 20) an[d] by Leonardo (Figure 34) show a position[er] that is readily dismantled and reerected. [In] addition to such a positioner, a winch (s[ee] Figure 21) was needed on the elevated construction site. If it was operated by four men, simultaneously, it had to wo[rk] long and hard to swing a three-ton load to the edge of the Cupola. For such a swing, see the perspective illustration ne[ar] the middle of Figure 26. Without a winc[h] the entire work force of Brunelleschi could not do this work, since a man pulling on a rope cannot, at any speed, li[ft] more than ca. 100 pounds.

[15] Nothing was changed with regard to vaulting without armature. As to other

contracts were made with ox drivers (Docs. 152–155) and a new rope was acquired (Docs. 138, 139).

Near the end of the fifth year some interesting parts were supplied by Donatello and Lorenzo:

12 October 1425. Donato di Nicholò di Betto Bardi, sculptor, shall have for his effort and mastery in making a *mozetto* for the hoist of the Great Tribune, weighing 29 pounds and 11 ounces . . . (calculation follows) . . . [and] Lorenzo di Bartolo, goldsmith, shall have for his mastery and effort in making five *mozetti* for the Opera, for the device of the Great Tribune, weighing 282 pounds and 2 ounces . . . of bronze . . . (calculation follows) (Docs. 140, 141).

Here, for once, there is reference to metal parts of the so-called screw vises. Even here, the record does not specify the screws, cooperating with these vises or nut blocks. Probably this omission reflects the fact that the blacksmith of the Opera made the screws of iron, while the matching bronze vise blocks had to be contracted for separately. Each positioning apparatus (Figures 18 to 21) required at least a pair of long screws, best made of iron, a vise block matching the screws, best made of bronze, and a wood frame or housing to provide the slideway.[16] It appears

that Filippo's friend as well as his rival contracted for the brass foundry and machining work needed to produce the screw vises. By then, the general success of the machine was well established, and only minor addenda or replacements were needed.

LATER MACHINES

The last six years of work in completion of the Cupola framework and walls, 1426–1432, no longer gave rise to significant changes in the crane system. The only document relating to a tangible part of this system shows that in December 1427 Filippo was authorized to "sell the big wheel" (Doc. 142). It is not stated whether this was a wheel of the ox hoist or scaffold and whether it was replaced or scrapped. The crane system continued to work, and the Opera made further contracts with ox drivers (Doc. 157 of 1431) and with other workers employed on the hoist (Doc. 249 of 1432). The Opera also settled its account with Filippo for the purchase of a horse. At some unspecified time he had used this horse "for [work on] the device, . . . then resold him and lost one florin." He was reimbursed for this loss (Doc. 156 of 1426).

The elevated parts of the hoisting system, as noted before, must have been readjusted as the masonry level reached higher points. We have noted some reflections of these readjustments and

---

methods and devices, the documents show gradual development, as noted here, but no drastic change at this particular time. See Appendix, pp. 141 f.

[16] If a slider of the type shown in Figures 18–21 had to support a weight of several tons, as it probably did, the iron screws extending through it had to have a diameter of at least about an inch, in which

---

case the casting for the slider was likely to weigh ca. 30 to 60 Florentine pounds. The estimate agrees well with Doc. 140 f. of 12 October 1427, cited in the text.

changes that are noticeable in the documents. It is mechanically obvious, although historically undocumented, that relatively large changes must have occurred when the framework and wall unit reached completion and when it became necessary to close it with an *oculus,* a stone structure weighing many hundreds of tons. This may have been the time when the use of the original crane system (Figure 26) was superseded by the first of the rotary or Lantern hoists (Figure 18). The latter hoist could stand on the uppermost platform of the original, central scaffold. It could act there as an upward extension of the scaffold and also as a replacement for the former load positioners. When this rotary crane had completed its function, probably during the years 1432–1434, it could be replaced by a similar machine or could be partly reassembled on a new platform supported by the *oculus* itself. This stage appears to be shown in the drawings by the San Gallos (Figure 32) and their contemporaries and successors. Thereafter, still higher cranes were needed and were supplied, as we have illustrated and briefly described (Figures 19, 27).

We may assume that Filippo used somewhat similar sequences of hoist installations in his other and later cupola constructions, the profiles of which are suggested in the scaffold drawing that forms a basic part of the technical record (Figure 8). It is not our intention at this point to describe these later constructions or the details of their design and actual building. However, utilizing the graphic and documentary evidence, we can approximately visualize the operations performed on such construction jobs, repeatedly witnessed by the Florentines and learned by the apprentice builders of the time.

A scaffold (Figure 8) was erected on the walls then standing and a hoisting machine (Figure 17) was built on the ground below it. When the load of a stone beam or other masonry material was ready for hoisting, it was attached to the free end of the hoisting rope by suitable hangers, while the other end of the rope was attached to the winch drum of the machine, or to one of the several drums if the machine provided for different operating speeds. Men or oxen were then applied to the machine, one or several at each end of the tiller. When they began to turn the drum, the load began to rise from the floor. When it arrived on top, no doubt a signal was shouted down and the driving effort was stopped. We may assume that workers up on the scaffold then attached an auxiliary rope (Figure 21) to the stone beam that had been elevated on the main rope. They pulled the stone beam out toward the place where it was needed, in which process they had to raise it considerably (Figure 26). They then used the triple hangers of the crane system (Figure 25), successively transferring the load from the several load-supporting screws of the main rope hangers to hanger screws of the load positioner. Thereafter the auxiliary rope could be detached, and the empty main rope could be allowed to swing back to the center. The load then hung from the vertical screw of

27. Lantern scaffold. Anonymous drawing, Florence, Uffizi, 248A. Photo Brogi-Anderson.

28. Brunelleschi's hoist. Taccola's version. Facsimile after Palat. 766, fo. 36ʳ, Biblioteca Nazionale, Florence.

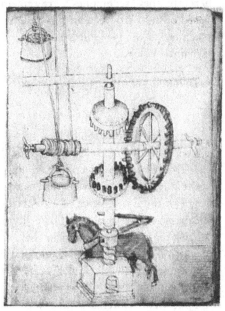

the positioner. By manipulation of the several screws the load could be placed accurately, for proper interconnection of stone beams and dovetails.

"Stones, brick, lime, sand, water, iron parts, wood parts, etc." were among the loads hoisted in the building of the Cupola (Doc. 148). No doubt the operation was similar in the construction of later cupolas. The sequence of purely manual and machine-assisted operations in elevated position could repeat itself, while the ox hoist on the ground continued to work.

For closing the Cupola of the cathedral, some special machine was used (Docs. 257, 260), followed by still other machines for the construction of the Lantern (Docs. 145, 274, 276). It does not appear that any one wrote a description of these techniques or that their details were remembered in the times of Manetti and Vasari. Inherently, these techniques were remembered only so long as scaffolds and machines of the types illustrated here remained in use.

OBLIVION OF BRUNELLESCHIAN MACHINES

It is not clear how long these machines remained in use. It seems possible that after Brunelleschi's time, building technique took a step backward as a result of the Renaissance of Vitruvian mechanics. The machinery of Filippo's and Lorenzo's time, shown in the *Zibaldone*, had been developed from workshop practices of the Trecento, but the machinery used soon after Filippo's and Lorenzo's time was developed—at least in large part—from ancient teachings, as a result of the Humanist trend of the times, shared by Filippo and Lorenzo. As a result, the more advanced machinery of medieval origin tended to fall into disuse. In the biographies of Filippo, this fact is reflected by the biographers' complaints about the poor care that the successors took of Filippo's models.[17]

One Antonio Manetti Ciaccheri was *capomaestro* of the Opera, some time after Filippo.[18] He had previously worked on the ox hoist, as carpenter, and had done model work (Docs. 249, 275, 304). The Lantern crane (Figure 18), or part of it, may be his design. It was a variant of the Brunelleschian devices, but apparently it was superseded in its turn by the subsequent and simpler Lantern scaffold, Figure 27. It is not clear whether still another variant, Figure 19, was used in the meantime, or whether this was only among the proposals made by competitors.

As we have shown, numerous artists made records of the cupola-building machines, without always attempting to describe them in detail or to record

[17] For example, see *Vasari's* Life of Baccio d'Agnolo. Regarding the lack of interest in Brunelleschian work in the time of Palladio, see *Sanpaolesi*, p. 135. However, we cannot agree when Sanpaolesi calls Nelli's work the superficial dissertation of an uninterested follower.

[18] The *capomaestri* from Filippo's to Michelangelo's time were Michelozzo di Bartolommeo, to 1452, Antonio Manetti Ciaccheri, to 1459; Bernardo di Matteo del Borra, to 1464; Tommaso Succhielli, to 1480; Giovanni di Montachuto, to 1507; Simone del Pollaiuolo, to 1508; and Baccio d'Agnolo, assisted by Giuliano da San Gallo, to 1514. *Cupola*, Docs. 290, 298, 301, 303, 336, 341, 342.

their cooperation and sequence. The drawings that Mariano Taccola and others made during Filippo's lifetime (for example, Figure 28) were collected, copied, repeated, or very slightly varied in a surprising number of portfolios. It is possible that actual use of the machines at the same time fell into oblivion.

The graphic record indicates only one major addition and change in the system, the gradual reappearance of mechanical concepts that had been described by Vitruvius. The renewal of such concepts occurred well after the times of Brunelleschi and Taccola. Lorenzo Ghiberti, during his last few years, tried to translate the ancient work, but did not achieve much more than a restatement of a few philosophizing introductions to the several books.[19] Alberti deciphered some more of the contents but hardly understood their more technical parts.[20] Francesco di Giorgio, probably assisted by Fra Giocondo or other humanists, produced a good part of a translation and thereby clarified some of Taccola's mechanics

and hydraulics.[21] At the same time he produced many variants, which merely reiterated or proliferated the earlier drawings and descriptions. Among them are miscellaneous variants of Brunelleschian machines, shown in Figure 29. The various drawings and texts of Francesco di Giorgio in turn were studied by copyists as well as capable engineers. The contents were even copied in marble reliefs (see Figure 30). Such copywork was useless for the progress of technology. Our impression is that the popular success of Francesco's *Trattati* led to the eclipse of the earlier machines built by Brunelleschi and described by Taccola.

Some Brunelleschian machines were of sufficient interest to ensure their survival. Remnants of load positioners or the like were preserved in one of the Exedras, at least until recently.[22] The San Gallos and others were able to produce realistic and detailed pictures of the devices (Figures 31, 32). In some cases they combined perspective

[19] Lorenzo Ghiberti, *Denkwürdigkeiten (I commentari)*, II, ed. J. von Schlosser, Berlin, 1912, pp. 13 f.

[20] Alberti complained that Vitruvius *ita scripserit ut non intelligamus* (VI, i, 91r). The many parts that were "not intelligible" to him he padded with erudition from Pliny and scholastic writers. As to late Quattrocento theories in this area see P. Fontana, "Osservazioni intorno ai rapporti di Vitruvio colla teorica della architettura del rinascimento," *Miscellanea di storia dell' arte in onore di I. B. Supino*, Florence, 1933, p. 309. Also see the editorial comment in *Trattati*, I, pp. XVII–XXIII, the review by R. Wittkower, *Architectural Principles in the Age of Humanism*, London, 1963, pp. 13 f., and the literature cited.

[21] In addition to Francesco's autograph translation of Vitruvius (Magliabecchiano II. I. 141, Biblioteca Nazionale, Florence), partial translations appear in the Saluzziano version of his *Trattati*, I, as follows: pp. 20 from Book III on cities; 36–39 from I on architecture; 39–44 from III on temples; 54 f. from V on theaters; 56–66 from IV on "the orders"; 67–70 from II on "the proportions"; 81–85 from VI on rooms; 100 from V on baths; 102–104 from VI on the Greek house; 105–107 from II on materials, etc. Other manuscripts of Francesco will be noted in connection with our chart of Brunelleschian influence, pp. 146 ff., *infra*. A typical, clarified restatement of a text by Taccola occurs in *Trattati*, I, pp. 180 f., based on Palat. 766, fols. 29, 34, 45.

[22] *Sanpaolesi*, pp. 109 f. Present whereabouts not reported.

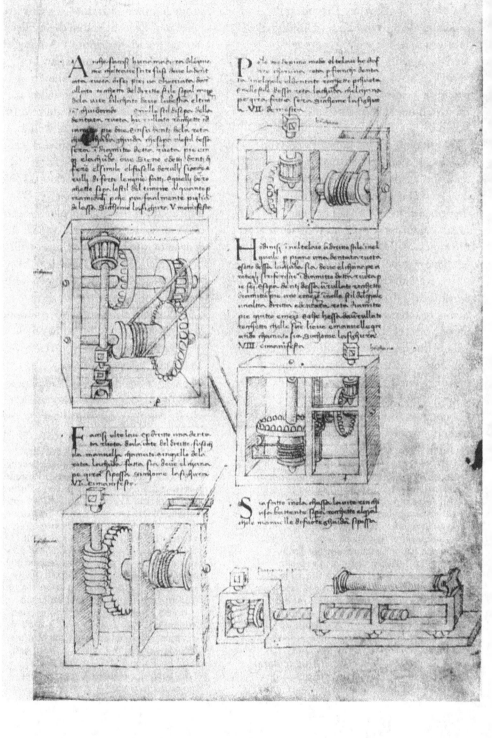

29. Variants of Brunelleschi's hoist or of Taccola's version, by Francesco di Giorgio or an assistant. Facsimile after Codex Saluzziano 148, fo. 50ʳ, Turin.

30. Load positioner and crane. Sculpture by a follower of Francesco di Giorgio, Ducal palace, Urbino. Facsimile after Baldi-Bianchini, *Memorie concernenti la città di Urbino*, Rome, 1724, Pl. LXXII.

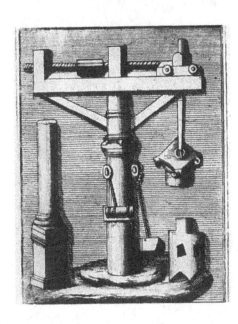

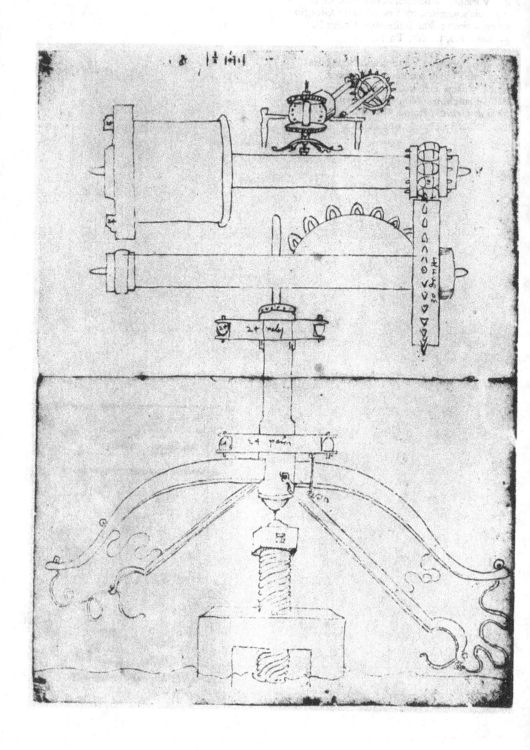

31. Parts of Brunelleschi's hoist, noted by
Giuliano da San Gallo. Facsimile from
Giuliano's *taccuino*, ed. R. Falb, Siena,
1902.

32. Elevated central crane, noted by
Giuliano, *ibid*.

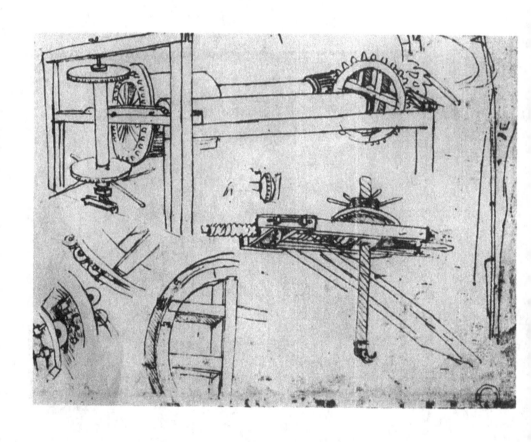

33. Brunelleschi's hoist and load positioner. Sketches of Leonardo da Vinci.
Facsimile after *Cod. Atlanticus* fo. 391ᵛ-b,
ed. Piumati.

showings with orthogonal projections.
Drawings of this type, made by
Giuliano da San Gallo, are so original
in technique and at the same time so
close to records kept by others—
draftsmen and notaries alike—as to
make it virtually certain that Giuliano
portrayed actual machinery that he
saw in Florence. It may be well to
remember that he was born only a
year before Filippo died.

Giuliano later worked in Rome, where
he was followed by his nephew and
pupil, Antonio the Younger. Between
the times of Bramante and Michelangelo, the San Gallos introduced
many elements of the Brunelleschian
tradition into Roman practice of the
sixteenth century. By the distribution
and use of many drawings, some of
them directly copied from such sources
as old Taccola, the tradition was transmitted more broadly into architectural
practice. The drawings and notes of
the San Gallos, largely preserved in the
Uffizi at Florence, show that specific
forms of Brunelleschian technology
were still active, although greater
numbers of these drawings and notes
are devoted to mechanical concepts
traceable to ancient sources.[23] We

[23] Filippo's hoist (Figure 17) seems unknown to Antonio, who shows only much
simpler machines. Occasionally he speaks
of a problem and does not seem to know
the solution, although Filippo had solved
it. For example, in Uffizi drawing A 1440ʳ
at left, Antonio shows a simple, man-
actuated hoist, copied from Taccola (Lat.
197, fo. 22ᵛ). Antonio adds a note saying
that an animal could actuate the machine,
but a man must then reverse the animal in
order to reverse the machine. It is as
though the machine of 1420 had never
existed.

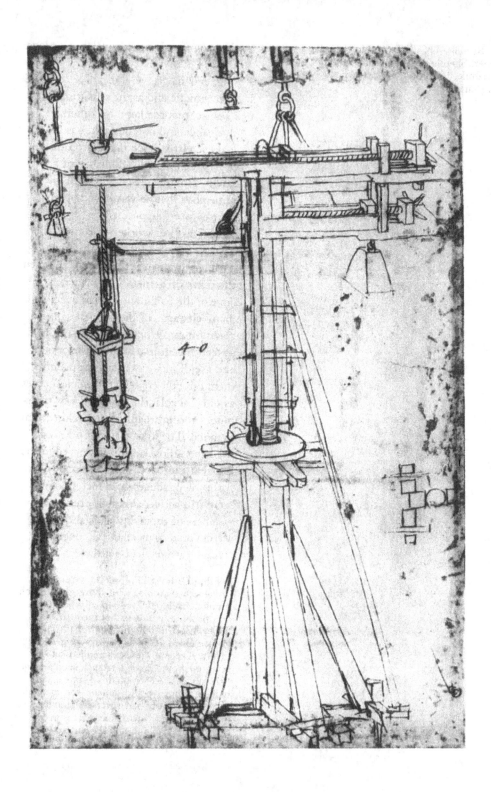

34. The load positioner. Additional sketch of Leonardo. *Ibid.*, fo. 349r-a.

mention this as a general impression. It is not possible at this point to explore these matters in greater detail.

The principal innovations of Brunelleschian mechanics, the screw-actuated reversing clutch and the screw-controlled load positioner, are unknown to the architecture books of Alberti, Serlio, Palladio, and their successors. The impression arises that these devices disappeared from building practice.

However, some of these and other innovations may have survived in other fields. The gear clutch reappears in a note book of Neroni, ca. 1550 (Figure 35), and in a printed work of Branca, 1629 (Figure 36). Several forms of the load positioner are copied in the notebooks of Francesco di Giorgio and his contemporaries and followers, including one of the several dozen marble reliefs of machines in Urbino, produced shortly before 1500 and elaborately shown in a later travel guide.[24]

In these later illustrations the machines reappear without reference to Brunelleschi, or to architecture. Apparently the late notes and pictures represent merely versions of copywork, done on paper or in other media but not in

[24] B. Baldi and F. Bianchini, *Memorie concernenti la città di Urbino*, Rome, 1724, Pl. 72, reproduced here as Figure 30. This particular variant of our Figures 20 and 21 had previously been copied in Francesco di Giorgio's *De architectura* (Harley 3281, London), fo. 12r, his *Trattati* (Saluzziano 148), Pl. 95, the London copybook Add. 34113, fo. 187r, and Leonardo's *Codex Atlanticus*, fols. 37v-b and 309r-b. As shown by Scaglia, *J.S.A.H.*, pp. 92, 113, it is also copied, exactly, in at least eight additional copybooks of the fifteenth and sixteenth centuries. Also see Besson, *Théâtre*, fols. 37, 38.

35. Brunelleschi's hoist (parts) in Sienese notebook S.IV.6 attributed to Bartolommeo Neroni. Facsimile from fo. 49r.

36. Variant of Brunelleschi's hoist. Facsimile from G. Branca, *Le machine*, Rome, 1629, fo. 21r of part I.

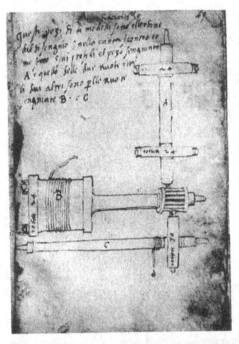

actual machine building, by followers of Francesco di Giorgio, Giuliano da San Gallo, and other epigones. However, even such modest copywork tends to preserve useful technical knowledge, which may, when once acquired, be adapted to new applications and possibilities. We think it will be justified in future studies to give further review to these technical copies and adaptations.

BRUNELLESCHI AS INVENTOR OF MACHINES

Brunelleschi intended his machinery for the Florentine Cupola and also for wider application, *tam super cupola maiori quam alio loco* (Doc. 124). No doubt he would have been pleased to know Branca's generalized application of his gear-reversal machine and would have regretted the demise of such machines in building practice. He may have led Florentine mechanics in new directions, but the question remains: Did he invent machines of general significance?

Surely his machines, as shown in the record considered here, amounted to more than the stone hangers admired by Manetti and Vasari. When one of the machines, such as the load positioner, was erroneously attributed to Leonardo, there were prompt exclamations about superior use of machine elements, as we have noted. However, a machine does not become historically significant because it is built by Leonardo, a great painter, or by Filippo, a great architect. If a machine invented by Filippo was a path-breaking innovation, one may expect to encounter its progeny in subsequent work, even if the invention be forgotten by the more philosophically inclined writers of architectural treatises.

The fact is that Filippo's complex machinery system (Figure 26) and even his individual machines (Figures 16, 17 and 20, 21) do not appear in subsequent building practice but only in notebooks and copybooks. Builders such as Antonio San Gallo returned to simpler methods. During the Cinquecento, Domenico Fontana had great popular success with his "Removal of the Vatican Obelisk."[25] The methods and machines that he used were elementary in comparison with Filippo's work. Nor did other mechanics, in Fontana's time, adopt mechanical systems comparable with those employed by Brunelleschi. Individual drawings of Mariano and his follower Francesco reappear in the Theaters of Machines, composed by the followers of Francesco.[26] Their development and even their plain reproduction no doubt served useful purposes. However, no one is known to have used or developed a mechanical system as advanced as that of Filippo, until the advent of the Industrial Revolution and its mechanized factories.

As might be expected under these circumstances, Brunelleschian machines

[25] D. Fontana, *Della trasportazione dell' obelisco vaticano*, Rome, 1590, *passim*, and C. Fontana, *Templum vaticanum*, pp. 109–173. See reproductions in Strauch, *History of Civil Engineering*, pp. 99 f. Also see Wittkower, *La cupola*, Pls. 32, 34, 37, 39a, 41.
[26] L. Reti, "Francesco di Giorgio Martini's Treatise . . . and its Plagiarists," *Technology and Culture*, IV, 1963, pp. 287–298.

are unknown to a historian of science and invention who wrote in Brunelleschi's century. The machines were noted by a Florentine historian of technology in the eighteenth century, but hardly beyond such limited extent as the machines had been described by Vasari.[27] Even Vasari is disregarded by modern historians of architecture and technology.[28] Even when other mechanical works of Gothic and Renaissance times were rediscovered, the character and achievement of the Brunelleschian machines remained forgotten.[29] Since then there has been no scarcity of interest in machinery concepts of various times, and there has always been enormous interest, for example, in Galileo's technology as well as his science.[30] Nevertheless, and in spite of Brunelleschi's incomparable fame, strangely the present studies are among the first attempts to review the

master's technical work, including his machine work, in some depth.

It is conceivable that Filippo was too far ahead of his time and therefore unable to influence his time with regard to machines. However, his system of hoisting and positioning devices hardly was reinvented even in later times. The progress of machinery, in those times, came about by studies in scientific technology, such as the mathematical and experimental analysis of gear-tooth profiles, lubricated bearings, and load-bearing posts or beams. These searching studies of simple things, not the developments of more complex devices, brought the major advances of modern technology. The fate of the Brunelleschian machines is comparable to that of refinements in Gothic tracery, forgotten in times of classic revival. It would be futile to insist that these machines were important to mechanical developments of later times. Perhaps they may be said to share the position of Filippo's method of vaulting without armature. In striking contrast to Filippo's structural and stylistic innovations, these famous machines and methods would now appear as matters of local and temporary sensation, not of major historic significance.

We are stating this in an effort to avoid misinterpretations of the mechanical work. We do not consider Filippo's machines as important inventions in the progress of machinery, but we consider them as most interesting elements in Filippo's progress. Some of his powers and also some of his limitations are shown with unequaled clarity

[27] Polydorus Vergilius, *De rerum inventoribus,* Venice, 1499; Manni, *De florentinis inventis,* pp. 79–83.
[28] T. Beck, *Beiträge zur Geschichte des Maschinenbaues,* Berlin, 1899; A. Wolf, *A History of Science, Technology and Philosophy in the Sixteenth and Seventeenth Centuries,* London, 1935; C. Singer et al., *A History of Technology,* II, III, Oxford, 1956–1957; A. Burstall, *A History of Mechanical Engineering,* Cambridge, Mass., 1965.
[29] Near the end of the eighteenth century, G. B. Venturi rediscovered the technical work of both Taccola and Leonardo. *Essai sur les ouvrages physico-mathématiques de Léonard de Vinci,* Paris, 1797. The publication led to a flood of popularizations of Leonardo.
[30] A general orientation is offered by L. White, "Pumps and Pendula: Galileo and Technology," *Galileo Reappraised,* ed. C. L. Golino, Berkeley, 1966.

in the record set forth by Ghiberti, the Opera, Taccola, San Gallo, and Manetti. For this reason the record is in need of further study. There may be manuscripts thus far undiscovered, which may throw further light on the development of Brunelleschian machines and related matters. We hardly think that all of these have been found, and we are sure they have not been evaluated in a final way.

So far as the image of Brunelleschi as machine builder and machine inventor is reflected by the record known to us today, it shows an architect-engineer who developed the mechanical elements then available into a highly refined system. He controlled the building and working of this system with iron willpower and rigid secrecy. By this remarkable *tour de force* he contributed little to the progress of machinery, but he contributed greatly to his personal success and that of his main work, the Cupola. As in his building method, which was free of various conventional expedients, he demonstrated once more the possibility of individual progress, beyond limits stipulated by ancient rules.

The biographers say that features of Roman ruins suggested to him the use of stone hanger tools, known in Roman times. It is possible that this was so, and it is even possible that he personally so informed Manetti. However, we think the hangers are secondary to his machinery, and his pertinent discoveries are secondary to the history of his mechanical conceptions and innovations. In a more significant way these innovations had nothing to do with Rome, Antiquity, and the Renaissance. They are part of the effort of a Gothic master, operating in his Gothic world. It may be for this reason that they were promptly forgotten. They were ingenious, but they belonged to a world then about to disappear, not to the world of his real successors or of his own major enterprises as architect.

BRUNELLESCHI AS PATENTEE AND CONTRACTOR[1]

6 One of the innovations claimed by Filippo Brunelleschi was a cargo ship. It led to one of his most curious ventures, wherein he appears against a background far removed from the Opera building yards. Here, we may see him as a protégé of Cosimo de' Medici, Florentine Consul of Maritime Affairs, as visitor to ancient offices in Pisa and in small towns along the Arno, and as one who desperately tries to save a cargo from a disintegrating craft. We see him also as recipient of one of the first patents of monopoly ever granted to anyone. The facts as presently known do not give us a clear picture of the technical nature of the ship, but the matter is full of interest for the economic historian as well as the student of Brunelleschiana.

The patent is not copied in the Opera documents published by Guasti, although the ship is mentioned in these documents. Neither the patent nor the ship is noted in Filippo's biographies, written by Manetti and Vasari. The document was discovered more than four hundred years after its date, by the Danish archivist Giovanni Gaye. After another century it became a matter of interest, first to historians of patents and then also to Brunelleschi's modern biographers.[2] We will review

it as an element of Florentine history and Brunelleschian historiography. Why did the government of Florence grant this patent to Filippo?[3] Why are Manetti and Vasari silent about it?

The patent reads as follows:

The Magnificent and Powerful Lords, Lords Magistrate and Standard Bearer of Justice,

Considering that the admirable Filippo Brunelleschi, a man of the most perspicacious intellect, industry and invention, a citizen of Florence, has invented some machine or kind of ship, by means of which he thinks he can easily, at any time, bring in any merchandise and load on the river Arno and on any other river or water, for less money than usual, and with several other benefits to merchants and others, and that he refuses to make such machine available to the public, in order that the fruit of his genius and skill may not be reaped by another without his will and consent; and that, if he enjoyed some prerogative concerning this, he would open up what he is hiding and would disclose it to all;

And desiring that this matter, so

---

[1] This is a revised edition of F. D. Prager, "Brunelleschi's Patent," *Journal of the Patent Office Society*, XXVIII, 1946, pp. 109 ff.
[2] G. Gaye, *Carteggio inedito d'artisti*, I, Florence, 1839, pp. 547 f., citing Archivio delle riformaggioni di Firenze, Provvis. Filza 113. Also see M. Frumkin, "Early History of Patents for Inventions," *Chambers Journal*, 1943, pp. 21 ff.; *Sanpaolesi*, pp. 60, 105 f.
[3] For the evaluation of inventions, the governments long contented themselves with inspection by more or less expert persons. The development of more significant methods began in France in the time of Richelieu. See F. D. Prager, "The Examination of Inventions from the Middle Ages to 1836," *Journal of the Patent Office Society*, XLVI, 1964, pp. 268 ff.

withheld and hidden without fruit, shall be brought to light to be of profit to both said Filippo and our whole country and others, and that some privilege be created for said Filippo as hereinafter described, so that he may be animated more fervently to even higher pursuits and stimulated to more subtle investigations,[4] they deliberated on 19 June 1421;

That no person alive, wherever born and of whatever status, dignity, quality and grade, shall dare or presume, within three years next following from the day when the present provision has been approved in the Council of Florence, to commit any of the following acts on the river Arno, any other river, stagnant water, swamp, or water running or existing in the territory of Florence: to have, hold or use in any manner, be it newly invented or made in new form, a machine or ship or other instrument designed to import or ship or transport on water any merchandise or any things or goods, except such ship or machine or instrument as they may have used until now for similar operations, or to ship or transport, or to have shipped or transported, any merchandise or goods on ships, machines or instruments for water transport other than such as were familiar and usual until now, and further that any such new or newly

shaped machine etc. shall be burned;

Provided however that the foregoing shall not be held to cover, and shall not apply to, any newly invented or newly shaped machine etc., designed to ship, transport or travel on water, which may be made by Filippo Brunelleschi or with his will and consent; also, that any merchandise, things or goods which may be shipped with such newly invented ships, within three years next following, shall be free from the imposition, requirement or levy of any new tax not previously imposed.

In order to determine how the invention and patent may fit into Filippo's story, we will begin by reviewing the place of this patent in the history of Florence, her constitution, and her economy.[5]

According to its own terms, the privilege was to be "approved in the Council of Florence." It is conceivable that approval was not forthcoming; however, this seems improbable, and even if it was so, it does not explain the biographers' silence. When the *signoria* once had acted, as it did here, Council approvals were mere matters of form. Filippo, with the support of the Opera, pursued the cargo ship enterprise for several years, actually disclosing his invention about 1424 and using it to

[4] *Audito . . . qualiter . . . Filippus . . . quoddam hedifitium seu navigii genus adinvenerit . . . et quod ipse tale hedifitium in publicum deducere recusat . . . et quod si aliqua prerogativa in hoc gauderet, quod celat apriret . . . deliberaverunt . . . .*

[5] We rely largely on Davidsohn, *Geschichte von Florenz*, II, ii and IV, i; A. Doren, *Italienische Wirtschaftsgeschichte*, I, Jena, 1934–1938, pp. 242–301; and R. von Pöhlmann, *Die Wirtschaftspolitik der Florentiner Renaissance*, Berlin, 1878, *passim*. Popularizations such as J. Burckhardt, *Die Kultur der Renaissance*, Basel, 1860, are no longer useful, except possibly for the history of nineteenth-century views.

secure an important contract in 1427. He had every opportunity to obtain an extension of the privilege, as he had access to the governing circles of Florence, and these circles governed the country by rules of economic expediency, not rigorous law. His father had belonged to the Major Guild of Judges and Notaries, and he himself was a member of another Major Guild, *Arte di Por San Maria*, the cartel for silk and gold.[6] He had access even to the *Parte Guelfa*, the feudal group—strangely considered as a gathering of citizen-merchants—which had control of most of the real estate and of substantially all *de facto* powers that even the Major Guilds did not possess.[7]

At the time of Filippo's patent these groups had reached the peak of their power as an oligarchy, which directed the constitutional and economic life of Florence. They continued to be "open" only to those who not only paid the matriculation fees,[8] but also had the approval of the rulers for their products or services: that is, "open" only for a small and self-perpetuating clique (including Filippo as a minor associate). This clique continued to maintain its own monopolies with great vigor, while purportedly fighting monopoly in general. The system gradually eliminated the monopolies and powers of the Minor Guilds.

The Major Guilds maintained church-building administrations, such as the cathedral Opera of *Arte della Lana*, the wool producers' cartel, and the Battistero Opera of *Calimala*, the wool importers' cartel. The guilds did not do this out of charitable contributions of their own but as a matter of administration of city taxes.[9] The economic rule of these guilds was entirely selfish. It was, to some extent, an enlightened rule. The population increased in numbers.[10] It would seem that it also increased in wealth. This was no longer the case when the Major Guild oligarchy was replaced by Medici rule, after 1434. The population then dwindled again, and the city, previously a capital of the cultured world, became more provincial. The lack of true democracy, which had been pronounced in the Middle Ages, became fatal.

Until Filippo's time Florence like other communes also had Minor Guilds, with some economic if not political significance, but they were in decay.[11]

[6] As to Filippo's goldsmith work of 1399, *Brunelleschiana*, p. 9, mentions work for master Benincasa Lotti, while *Sanpaolesi*, p. 13, mentions master Lionardo di Matteo Ducci. Also see Prager, "Brunelleschi's Clock?" p. 209. Some writers assume that he previously followed a juridical study, but there is no authentic proof of it.

[7] Davidsohn, *Geschichte*, II, ii, pp. 117 ff.; IV, 1, pp. 103 ff., 184 ff., IV, ii, pp. 1 ff. Also see *Vasari*, pp. 232 ff.

[8] Ten to 20 florins (Pòhlmann, *Wirtschaftspolitik*, pp. 46, 50–63, 140, and statistical appendix).

[9] A. Doren, *Die Florentiner Wollentuchindustrie*, Munich, 1901, pp. 201 ff., and *Das Florentiner Zunftwesen*, Munich, 1908, pp. 53 f., 703 f.

[10] R. Davidsohn, "Blüte und Niedergang der Florentiner Tuchindustrie," *Zeitschrift fur die gesamte Staatswissenschaft*, LXXXV, 1928, pp. 225 ff.

[11] G. Renard, *Histoire du travail à Florence*, II, Paris, 1913, pp. 19 f.; Doren, *Zunftwesen*, pp. 121 ff.; Davidsohn, *Geschichte*, IV, ii, pp. 26 f.; Holt, *Documentary History*, I, pp. 102 ff. Confused, romanticizing accounts are often given,

Filippo himself, by demonstrating total disregard of the Guild of Masters of Stone and Wood on many occasions, promoted the process whereby this group lost even the last remnant of power over its field, that is, over sculpture, architecture, masonry, and carpentry. The exclusive right of this guild had been narrowed greatly in 1325 by regulations that the Major Guild government then imposed, and such powers were all but abolished in later decades of the Trecento under the oligarchy of the plutocrats. Filippo seemed to disregard the guild. The masters tried at one time to rebel against him, but failed to achieve anything.[12] Even more simple was the entry into the trade of boatmen, as this trade had no Florentine guild at all. Filippo could enter without opposition by competitors and apparently did so simply by floating his newly invented ship. The general trade in this area was protected only by a saving clause, which the state imposed on Filippo's privilege, in favor of those who wished to continue to use "such ship or machine or instrument as they may have used until now."[13]

Having entered into this trade, Filippo had to deal with a power that was stronger than a minor guild: his city's erratic river. The Arno, which connects Florence with Pisa and the sea, is very treacherous. Only for some months of the spring and early summer does it support a certain amount of traffic.[14] Some boats had been plying the river since Antiquity, but in the Middle Ages this traffic already seemed inadequate for the two cities and their enterprises, particularly in the absence of a unified administration but also later when such administration was in effect.[15]

The Florentine Opera had decreed in 1319 "that the people are to lend a helping hand when marble is shipped from Pisa to Signa in cargo boats."[16] The traffic was conducted by small boat operators, to whom the Opera gave "security and license for their services." Thus the operators could expect Florentine police protection and payment for goods unloaded.[17] In 1334, the Opera decided to acquire its own boats,[18] but private operators continued to appear. They brought a variety of

---

for example, by J. E. Staley, *The Guilds of Florence*, London, 1906, pp. 58 f.

[12] *Cupola*, Docs. 116–118; *Vasari*, pp. 218 f.

[13] In other documents, clauses of this type were imposed at the request of guilds, as shown by F. D. Prager, "A History of Intellectual Property," *Journal of the Patent Office Society*, XXVI, 1944, pp. 711 ff.; XXXIV, 1952, pp. 106 ff. As also shown in these articles, precedents for the clauses go far back into the Middle Ages.

[14] It has ca. 145-foot rise in ca. 50 miles from Pisa to Florence. About its course in Brunelleschi's time, see the maps in G. Guarnieri, *Da Porto Pisano a Livorno Città*, Pisa, 1967, pp. 10, 30, 102, 130.

[15] A. Schaube, *Handelsgeschichte der romanischen Völker*, 1906, pp. 55, 654 ff.; Doren, *Wirtschaftsgeschichte*, p. 382; Davidsohn, *Geschichte*, IV, ii, pp. 27, 270 ff.; J. W. Brown, *The Builders of Florence*, 1907, pp. 2 ff., 9 ff. For similar medieval traffic on other rivers, see Holt, *Documentary History*, pp. 10 f., 56.

[16] *Costruzione*, Doc. 30; Davidsohn, *loc. cit.*

[17] *Costruzione*, Doc. 62 (year 1343): . . . *concedimus . . . nautis . . . plactarum . . . qui . . . conducerent marmorem . . . pro constructione matricis ecclesie . . . quatenus . . . in veniendo . . . et redeundo . . . sint liberi et securi : . . . post opportunam inspectionem.*

[18] *Costruzione*, Doc. 43.

"cargo vessels, boats, and other navigable crafts," probably moving upstream by rowing or pole-setting.[19]

At later times the boats from Pisa may have landed in Florence, which would indicate some late Trecento river clearance, otherwise unreported, but they continued to arrive in very irregular sequence. If the published Opera documents are complete, there were five such arrivals in February 1354, two in April, and then none for a long time. All of these, noted in the Opera books, brought small columns and smaller pieces of marble. There was a schedule of prices and fees, as was usual. In June 1355, the Opera hoped "to find a method of obtaining the most needed marble from Pisa," and again near the end of 1356, it expected "a new shipment of marble," but it is not clear that the shipments arrived.[20] A good part of the problem was due to hostile actions of the Pisans and Florentines.

In Pisa the technology of such transports was well developed,[21] but the administration had the defects of medieval planned economy. *A Consolato del mare*, sometimes called *Ordo maris*, inspected all boats, all devices and merchandise thereon, and all pertinent contracts, earnings, and activities of "boatmen who carry stones on boats," as well as it tried to regulate the lives of all. This office issued a special permit for each trip up the Arno.[22] It was a case of total state control, the equivalent of state monopoly with a system of licenses. The practice stemmed from traditions of regions farther south and east. It then became the practice of Florence, as one of the first effects of Medici rule. In Florence, as elsewhere, it soon became a total failure, as may be expected of any economic system that allows even less freedom than was practiced under the guild system.

During the first two decades of the Quattrocento, the city of Pisa and her possessions, including Livorno, fell to Florence. Some Florentines, and especially Cosimo de' Medici, then hoped for larger and more profitable navigation, particularly on their Arno as well as overseas. Their first major action in pursuance of this hope was not a constructive one. It was a decree which in effect expropriated the wealthy Pisans and created a Florentine state monopoly for navigation, *de facto* transferring the monopoly of the Pisans to Florence.[23] Meanwhile, as mentioned, the monopolies formerly held by Florentine guilds were in the process of disappearing. However, a new institution ap-

[19] Propulsion by oars is mentioned by G. P. Pagnini, *Della Decima . . . e della mercatura de'fiorentini*, II, Florence, 1765, pp. 212 ff. Pole-setting is mentioned by Staley, *Guilds of Florence*, p. 64.
[20] *Costruzione*, Doc. 70 at pp. 80, 82, 87 f., 90 ff.; for a schedule of prices see *Costruzione*, Doc. 68.
[21] See Rodolico, *Le pietre*, pp. 247, regarding Cornetto's and Buschetto's transports from Elba and Sardinia for such buildings as the Pisan Dome.

[22] A Schaube, *Das Konsulat des Meeres in Pisa*, 1888, pp. 30–101, citing sources dating since A.D. 1200.
[23] Schaube, *Das Konsulat*, pp. 213 ff.; Pöhlmann, *Wirtschaftspolitik*, pp. 49, 68 ff., 102; C. S. Gutkind, *Cosimo de' Medici*, London, 1938, p. 179. Also see *Vasari*, pp. 224–227, about Brunelleschi's relation to Cosimo.

peared: an inventor's private right—state-recognized, exclusive or monopoly—to his invention, for some years' duration. Forerunners of the institution are rare, and in Florence the practice remained rare also in later times. Innovators generally received only tax exemptions and the like.[24]

The date of Filippo's ship invention is unknown. It may lie months, or more probably years, prior to the month of June 1421, when the *signoria* stated that he was "hiding a secret." The idea of a transport ship appears secondary to the hoisting methods that Filippo studied in the early years. Perhaps we may estimate that the ship idea came to him at some time between 1415 (when he may have traveled to Pisa)[25] and 1421.

The form, construction, and equipment of Filippo's ship are also unknown.[26] The patent only asserts that he could transport loads "for less money than usual." Certain drawings (Figure 37) show roller structures to facilitate loading and unloading of columns; perhaps this was the innovation or one of the innovations.

The boat was built and was known as "Il Badalone," meaning a simple-minded and wayward giant. The word may also relate to *battello* or ship. Whether the vessel was able, or even intended, to travel over the sea can be doubted.

There were long delays in completing the Badalone, and the affair became a source of irritation. The original three-year term of the patent had long expired before the ship was finished; it is conceivable that Filippo tried to extend this time by negotiation with the government, but such negotiation by itself is irritating. One of Filippo's deputies and rivals at the Opera was Giovanni di Gherardo from Prato, called Acquettini, a lecturer at the *studio* of Florence.[27] He not only opposed Filippo's Cupola design, because of the darkness that would prevail under it, but he also published a sonnet attacking his opponent in general. He ridiculed the ship in particular, using the coarse invective and heavy sarcasm that was usual at the time. Filippo replied in kind. Neither of them produced a masterwork of elegant writing, but the exchange of sonnets is historically interesting. It reveals the rivals' mode of expression, and to some extent their mode of thought, in a more direct form than we find in the notarial documents of the Opera or the engineering drawings made by various rivals and followers.

[24] Prager, "History of Intellectual Property," XXXIV, pp. 111–137; H. Silberstein, *Erfindungsschutz und merkantilistische Gewerbeprivilegien*, Zürich, 1961, pp. 1–25, 111–123, 144–166; Davidsohn, *Geschichte*, IV, ii, p. 12. Occasionally the government granted combined copyrights and patents to authors of technical books, such as F. Veranzio, *Machinae novae*, ed. Florence, 1615; see H. T. Horwitz, "Ein toskanisches Privileg," *Technik und Kultur*, XIX, 1926, pp. 202 ff.
[25] *Fabriczy*, in the chronological appendix.
[26] Frumkin, "Early History" and *Sanpaolesi*, pp. 117 f., would provide the ship with devices, especially screw-acting ones, for loading purposes. We are unaware of a basis for this postulate.

[27] The following is based, in part, on Guasti, *Belle arti*, pp. 109–128. About the *studio* see A. Gherardi, *Statuti della università e studio fiorentino (1387)*, in *Documenti di storia italiana*, VII, 1881; Davidsohn, *Geschichte*, IV, iii, pp. 142–146; Gutkind, *Cosimo*, pp. 73–75, 229–231.

37. Marble transport according to Taccola. Facsimile after Palat. 766, fo. 40ᵛ–41ʳ.

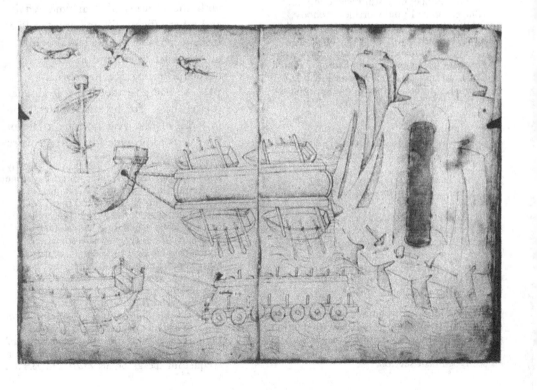

Even the sonnets are enigmatic. They hardly give us the exact terms used by their authors; the handwritten texts appear to be garbled at several points. Nor is it clear to what extent, or under what circumstances, the exchange took place in public.[28] Different readers will have different views about the intellectual achievements embodied in the verses. However, in any case the basic meaning of the sonnets is clear. It is capable of fairly literal translation:

Sonnet of Giovanni Acquettini from Prato to Pippo di Ser Brunellesco

O you deep fountain, pit of ignorance,
  You miserable beast and imbecile,
  Who thinks uncertain things can be made
    visible:
There is no substance to your alchemy.
The fickle mob, eternally deceived
  In all its hope, may still believe in you,
  But never will you, worthless nobody,
Make that come true which is impossible.
So if the Badalon, your water bird,
  Were ever finished—which can never be—
  I would no longer read on Dante at school
But finish my existence with my hand.
  For surely you are mad. You hardly know
  Your own profession. Leave us, please,
    alone.

Sonnet in Response to the Preceding One, Made by Pippo di Ser Brunellesco

When hope is given us by Heaven,
  O you ridiculous-looking beast,
  We rise above corruptible matter
And gain the strength of clearest sight.
A fool will lose what hope he has,
  For all experience disappoints him.
  For wise men nothing that exists
Remains unseen; they do not share
The idle dreams of would-be scholars.
  Only the artist, not the fool

Discovers that which nature hides.
Therefore untangle the web of your verses,
Lest they strike sour notes in the dance
When your "impossible" comes to pass.

These mutual invectives may be pertinent to the Brunelleschi story in more than one regard, but certainly they show the tenor of a debate about the Badalone.[29] We have here Acquettini's offer and Brunelleschi's rejection of a foolish wager: failure of the ship against failure of Acquettini. In fact, both failed.

Acquettini must have written his lines before the end of 1425, since he in that year—when Filippo was "Lord Magistrate"—actually lost his job as recitalist of Dante. Filippo's preparations continued. In 1426, he obtained two leaves of absence from the Opera to negotiate with the Consuls of Maritime Affairs and other offices, still operating in Pisa. It seems possible that he had the ship built in that town, and in any case he needed a loading site. In February 1427, he had a further leave of absence for ten days "for his own affairs," and in April of that year he took a trip of four days to provide for "certain work

[28] It may have been limited to the circle of Cosimo de' Medici, as was the production of sonnets about San Lorenzo (*Vasari*, p. 226; also see *Manetti*, pp. 66, 70). Text: see Document section.

[29] Guasti, *Belle arti*, pp. 109 ff., followed by *Fabriczy*, pp. 390–393, interpreted Badalone in Acquettini's sonnet as meaning (also?) the Cupola. To us it seems to mean primarily if not exclusively the ship, which has the same name in *Cupola*, Docs. 110, 112, and is clearly alluded to when Acquettini speaks of an aquatic animal, *che 'n acque vola*. Both sonnets may allude to studies of perspective or to debates about illumination inside the Cupola, when they refer to attempts to make the invisible visible, *lo 'ncierto altrui mostrar visibile*. Brunelleschi's position, both as member of the silk manufacturers' guild and as builder, may be meant when Acquettini speaks of his *ordir* and *tessere*.

of his own."[30] The latter work was believed to redound to the honor of Florence; on this occasion he also obtained the loan of a long rope, perhaps indicating that his ship was to be towed.[31] On 7 May the Opera wrote a letter for him to the *potesta* of Castelfranco on the lower Arno, asking him "to give aid, counsel and favor to Filippo, as he may require." In Castelfranco, shipping franchises were issued, which may have been needed in addition to those issued in Pisa.[32] The progress of a merchant-adventurer and inventor was not easy.

Filippo then proposed to ship marble lying in Pisa to the Opera in Florence, an operation similar to previous contracts made by him and other architects.[33] The Opera noted

12 June 1427. The administrator may contract with Filippo di Ser Brunellesco for the shipment of 100 tons (*migliaia*) of white marble from Pisa to Florence, entirely at his expense, for 4 lire, 14 soldi per ton. (Doc. 107)

The price was less than one-half of that paid somewhat later for marble delivered by oxcart and also was below the price for marble delivered in small boats.[34] Filippo obtained advance payments of parts of the purchase price:

14 May 1427. To Filippo di Ser Brunellesco, 15 Florins, allotted for a certain quantity of marble that he shall convey to the Opera, from the supply at Pisa (Doc. 106).

12 June 1427. 40 Florins, loaned to him for the delivery of 100 tons of marble that he shall bring from Pisa to the Opera (Doc. 108).

Thereafter, the documents are silent for a year; then comes evidence that the Badalone made part of a trip from Pisa toward Florence (probably in the autumn, winter or spring, when the water was high) but that it reached only Castelfranco, about halfway up the river, or Empoli, another ten miles up. The Opera notified Filippo that he

. . . shall be required within eight days next following to ship by small boats to the Opera that quantity of white marble which he had shipped on the Badalone from the city of Pisa to Empoli and Castelfranco. Failing this, the administrator shall have said marble shipped by small boats to the Opera within the present month (Doc. 110, 12 May 1428).[35]

It is clear that Brunelleschi had been forced to unload marble prematurely, either at a point between Castelfranco and Empoli, or partly at one town and partly at the other. The next document,

30 He was "required" to visit the offices in Pisa (*Cupola*, Doc. 97 ff.).

31 *Cupola*, Doc. 100: *unus canapis;* Doc. 111: *uno canapo che avevangli prestato, di lib. 240.* It is not stated how Filippo transported this weight of 240 Florentine pounds. If the rope was 1 inch thick, it was over 500 feet long.

32 *Cupola*, Doc. 105; E. Repetti, *Dizionario geografico fisico storico della Toscana,* Florence, 1833–1843, s.v. Castelfranco.

33 For example, by Ghini, Fioravante, and their successors. *Costruzione*, Doc. 70, at pp. 95 ff., 104 ff., 111 ff. A typical provision (*Costruzione*, Doc. 72 at p. 121) was *Faciala a sue spese se non viene ben fatto, e se viene [ben] fatto, avrà provisione.*

34 *Cupola*, Docs. 159, 164.

35 . . . *infra otto dies proxime futuros teneatur conduci facere illam quantitatem marmoris albi quam conduci fecit a civitate Pisarum usque ad castrum Empoli et Castri Franchi cum Badalone, cum schafris usque ad Operam . . . .*

dated four and a half years later, shows that he had hoped to reach Florence with his ship:

> 3 December 1432. They have entered to the charge of Filippo di Ser Brunellesco the contract that he has with the Opera: to deliver from the city of Pisa to the Opera, at his expense and with his vessel called the Badalone, 100 tons of white marble, at 4 lire, 14 soldi per ton (Doc. 112).

Evidently none of the marble had been delivered by then. It had been lost "in the Arno."[36] Thus it seems that the Badalone had sunk or disintegrated, although the document of 1432, some five years after the trip, speaks of the ship in the present tense. It is not clear whether any of the marble ever reached Florence. We read after another half year, in June 1433, that three contractors shall deliver a total of 600 tons of marble by October 1433. The Opera agreed to accept this marble at 7 lire, 10 soldi per ton if it arrived on board small boats, or at 9 lire, 16 soldi—more than twice the original price—if it came by land (Doc. 164). No further trip of the Badalone is recorded.

Brunelleschi had lost his investment in the ship and probably a large part of the cost of the marble. His taxable assets were about 3,000 florins in 1431 and only a little more than 2,000 two years later, when the Badalone disaster had been paid off. During the next ten years his assets rose again to the former level. Evidently the patent and marble-contracting venture was a bad failure, but Filippo took it in his stride. A few

years later he acted as contractor again.[37] He was not easily subdued by such adversities as the shallowness of the Arno, the poor development of mechanized floats or ships, and the complexities of contracts.

Why, then, was the affair so totally disregarded by Manetti and Vasari? Surely it was full of interesting human sidelights, which biographers generally love. Were the writers confused about the part played by Cosimo de' Medici, or embarrassed about it? Both explanations are possible. The biographers are similarly reticent about another Medicean-Brunelleschian failure, the campaign against Lucca in 1430, when Filippo tried to "convert the city into an island" but only converted the Florentine camp into a lake.[38] However, was this a reason for withholding from the biographies all reference to the patent document, which had high official praise for Filippo's inventiveness?

It is conceivable that the document and the affair were forgotten by the time his biographies were written, except that the Florentines were not forgetful in general. In the present case there is good evidence that the affair was remembered. As in other instances,

---

[36] This expression occurs in Filippo's tax return of 1431: *Fabriczy*, pp. 510 ff.

[37] *Fabriczy*, pp. 510 ff.; *Brunelleschiana*, pp. 25 ff. (tiles for the Tribunes, 1436).

[38] *Vasari*, p. 227, mentions the war but not the part played by either Cosimo or Filippo. The whole affair was one of the anti-Ghibelline actions that had been frequent in the *Comune* and were no longer popular under the *ducato*. Lucca belonged to the Empire. Florence, allied with the Pope, attacked it, as it had on former occasions. The planning was confused. Also see the next following study, note 7.

the evidence comes from Mariano Taccola (1382–ca. 1454) and his copyists. In 1433, when the Badalone enterprise was in its final stage, Taccola completed a booklet on unusual engines and devices (Palat. 766), wherein he showed several of Brunelleschi's inventions and developments. Folios 40ᵛ–41ʳ, shown here as Figure 37, show a marble transport by ship, and the text on folio 41ᵛ shows clear allusions to the Brunelleschian experience.

It begins with instructions about how to quarry marble blanks for columns and to "let them down to earth." Taccola then refers to a special windlass, shown on his text page, and an equally special "wagon with fourteen wheels" also used as a float and pulled by a ship.[39] The first part of the text is detailed and slow-moving, but it gathers speed, and then breaks off:

> . . . Then have vessels in the sea, near the strand, with capstans to draw the said column. [One capstan or windlass is shown on the verso of the second drawing page, below this text.] Make the vessels firm by their anchors. [Then, without transition, the tenor of the text is changed.] Let it be known that one cannot explain each and every detail. Ingenuity resides in the mind and intelligence of the architect rather than in drawing and writing. Many things occur in the course of the work that the architect or worker never planned. Therefore let the architect be experienced and learned, mindful of much he has read

and seen, and always prepared for action. One thing however is mainly to be noted: if nature has not given subtle and perspicacious genius to an architect, he is worth little. Only if nature has given him power over that which fortune may bring (*si dotatus a natura cum accidentale aquisito*), can he boldly exercise his art.

In general outline this text is an abbreviated chapter from Pliny the Elder, which was also used by Ghiberti in his memoirs, although in very different ways.[40] Ghiberti derived some aesthetic views from Pliny. Taccola, like the ancient author, discusses marble quarrying, marble transport, and general aspects of the architect's work, in this sequence. His discussion is laconic, as usual, while Pliny's text has many asides.[41] In particular, Pliny finds numerous occasions to insert remarks about luxury; Taccola omits them. Taccola's tone in this text also differs from his usual one, which almost invariably is dry and factual. Here he displays abruptness and apparent irritation in introducing the architect's reaction to certain questions about his methods and the exact apparatus used for each purpose. The prose style appears to contain the influence of another personality. We think the personality reflected in these statements is Brunelleschi.

[39] Such wagon floats (also mentioned in Palat. 766 at fols. 53ᵛ–54ᵛ) were of interest in connection with the work of divers.

[40] Pliny, *Historia naturalis*, XXXVI, 1–21; Ghiberti, *Denkwürdigkeiten* (*I commentari*), II, pp. 14 f.
[41] Pliny, Book XXXVI, 1–3, speaks of marble cutting and marble transport; 4–10, of sculptors and stone-cutting techniques; 11–13, of types of marble; 14–17, of architectural achievements and "wonders of the world."

One could visualize Filippo and Taccola talking about marble shipment, among other mechanical-hydraulic operations supervised by the *architet-tore*. Taccola may have translated "Plinio" for his famous visitor. It further appears that the visitor became hard and sharp when he was reminded of questions raised about a marble shipping procedure. The talk could have taken place between 1425, when the Badalone enterprise became controversial, and 1433, when Taccola's book was finished.

Taccola's book was popular. The illustration showing the marble transport vessels was frequently copied or developed, and occasionally the text was translated. A drawing fairly similar to the present Figure 37 may also be found in the work of Lorini, one of the early printed "Theaters of Machines," the contents of which are largely derived from the work of followers of Taccola's technology.[42] Thus it appears that artist-artisans of the later Renaissance knew the Brunelleschi-Taccola version of the ancient lore of marble transports. After the exchange of sonnets and the ensuing momentous failure of the Badalone, it would be strange if the Florentines did not remember something of the venture

that was connected with this development.

Nevertheless, as mentioned, the biographers say nothing about it. In Manetti's case the silence may be explained by the fact that his booklet remained unfinished, but Vasari's work is finished and contains a number of Brunelleschian episodes not found in Manetti. Why does this major episode not appear? Why is there not at least some part of it, as in the case of the episode of 1430 at Lucca? We could hardly assume that the affair was too small to attract the biographers' interest. They wrote at length about much smaller matters, such as the stone hangers of the cupola hoist. Evidently the story was suppressed, perhaps because it embarrassed the biographers.

If so, the reason probably lies in the transformation of values that came with the transition from the Guild oligarchy to the Medici monarchy or from the world of artisan-artists to that of professional intellectuals and specialist performers. Manetti, influenced by the Medici system, and Vasari, totally formed by it, no longer understood Brunelleschi's time on its own terms; they understood it only with moralizing eclecticism of their own times. Brunelleschi's power came partly from the fact that the restraints of the Middle Ages disappeared in his time, and those of later times had not yet appeared in full. This enabled him to "deal" freely in many materials, including silk, gold, jewelry, bronze, iron, wood, paint, rope, stone, brick, mortar, and the brawn of pole-setting boatmen or of towing animals. In the times of Manetti

[42] Quattrocento translations include Palat. 767, p. 144 f., Ital. Z 86, fols. 49v–50r; and Add. 34113, fols. 30v–31r, 31v. Also see B. Lorini, *Le fortificazioni*, Venice, 1609, p. 191. Between the times of Taccola and Lorini, Antonio San Gallo the Younger had shown a similar scene in his drawing, Uffizi, Arch. 794A. Also see J. Leupold, *Schauplatz der Hebzeuge*, Leipzig, 1724, p. 45, and Pl. IX, Fig. VI.

and Vasari, versatility was still admired, but the versatility of Brunelleschi no longer had full acceptance. Even his studies of clockwork received only passing attention, and his activities as a contractor or cargo-boat operator probably would have been mentioned with excuses even if they had been successful. These writers also disregarded the commercial operations of Ghiberti's foundry shop or noted them only with embarrassment and even a critical attitude. The net effect was shallow moralizing about general goodness, supposedly shown by one of the artistic heroes. Vasari even purified the images of his heroes between the times he wrote the first and second editions of the *Lives*. The academic idea of "pure art" began to grip the popular mind.

This idea, or illusion, became more dominant during some periods of modern times. Fabriczy, in his generally well-reasoned work on Brunelleschi, found it necessary to criticize the master for being so selfish as to accept the monopoly patent, the existence of which had been noted by Gaye. Apparently Fabriczy would have wished to see "the artist" unaffected by some of the human impulses felt by artists of the early Quattrocento. We must attempt to see Filippo without such prejudice, even where we see him operate in a field far removed from the activities of modern artists.

BRUNELLESCHI ON INVENTION AND BUILDING[1]

7 A Brunelleschian statement is recorded with unique directness in one of the books of Mariano Taccola, "the Archimedes of Siena." This notary, sculptor, and engineer produced an illustrated treatise, De machinis, which became famous when some copies of it were rediscovered about two hundred years ago, and he also wrote a set of smaller books or drafts constituting another treatise, De ingeneis, and a set of less organized drawings, descriptions, and autobiographic remarks.[2] When some of these papers were discovered, near the end of the nineteenth century, it was observed in passing that they contain among other things a speech given by Brunelleschi in Siena.[3] Speeches are also inserted in Vasari's Life of Brunelleschi, but they are fictitious dramatizations in the style of Thucydides. By contrast, Taccola writes down what he personally heard and only translates it into his notarial Latin. We will presently note clear evidence that the reported speech alludes to facts that are confirmed by Brunelleschian documents. These do not occur in the popular reports of the later biographers, and Taccola could have had them from only the master himself.

The speech was overlooked by the earlier literature on Brunelleschi. The reason is that the speech, like many other parts of the early notebooks, had not been transcribed and published.[4] These are now beginning to find more interest and are indeed entitled to close review. They add direct and most authentically stated facts to the material contained in the biographies, the Opera documents, and the engineering notebooks. It happens rarely that the personal comments of a great artist are recorded by a notary, but it happened here. Of course even a notarial record resembles a biographer's report in that it gives only one man's written testimony about things that others said or did. Taccola's book is uniquely vivid, as it is written in an expressive, medieval style and illustrated by quaint and spontaneous sketches. The writing seems to imitate Brunelleschi's own language by specially animated and even flamboyant expressions. The pertinent pages are shown here (Figure 38), a transcription of the whole is in

[1] This is a revised edition of F. D. Prager, "A Manuscript of Taccola, Quoting Brunelleschi, on Problems of Inventors and Builders," Proceedings, American Philosophical Society, CXII, 1968, pp. 131 ff.
[2] We have discovered that the manuscript Lat. 28,800, now in Munich, is the autograph of De machinis. It is about to be published. As to earlier literature about De machinis and its copies, see L. Thorndike, "Marianus Jacobus Taccola," Archives internationales d'histoire des sciences, VIII, 1955, pp. 7 ff. De ingeneis comprises Lat. 197, Munich (Books I and II and a Sequel) and Palat. 766, Florence (Books III and IV).
[3] Jähns, Kriegswissenschaften, p. 278; Thorndike, "Marianus," pp. 7 ff. Also see P. L. Rose, "The Taccola Manuscripts," Physis, X, 1968, pp. 337–346.
[4] The best publication of drawings from this manuscript is M. Berthelot, "Pour l'histoire des arts mécaniques et de l'artillerie vers la fin du moyen âge," Annales de chimie et de physique, 6e série, XXIV, 1891, pp. 433 ff., especially pp. 472–496.

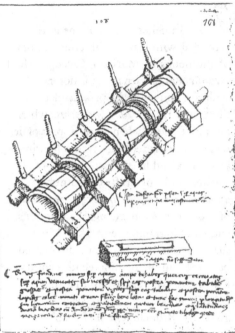

38. Brunelleschi's speech; riverworks and harbors, after Taccola. Facsimile from Lat. 197, fo. 107ᵛ–109ʳ, Munich, Bayerische Staatsbibliothek.

our Document section, and we will give a translation.[5]

From facts reported by Brunelleschi's biographers, it is fairly easy to recognize an occasion for his meeting with Taccola. He traveled often to Rome, and Siena is on the way. In addition, Taccola was his colleague in more than one sense. Both had begun as sculptors and had become engineers and inventors. The Sienese also designates himself as a writer, a miniaturist, and a man able to take care of waterworks.[6] Both men were interested in constructions dealing with water. In Taccola's books there are many ships, bridges, flood controls, foundations submerged in water, aqueducts, tunnels under mountains, and (following an ancient hope and illusion) siphons over mountains. In Brunelleschi's case it will suffice at this point to mention his ship venture and his later dam-building venture at Lucca.[7]

Taccola may have met Brunelleschi, or at least seen some of his machines, before Brunelleschi came to visit him. The approximate dates of the visit and of Taccola's report can be determined. It is possible to trace the evolution of Taccola's handwriting and drawing

[5] The manuscript, as shown by the facsimile, records consecutive parts of the speech on two verso pages, while beginning a related discussion of technical character on the opposite recto pages. This is one of many editorial innovations and oddities in Taccola's work.

[6] Palat. 766, fo. 69ʳ, partly transcribed by L. Gentile, *I codici palatini della reale biblioteca nazionale di Firenze*, II, Rome, 1885, p. 296.

[7] About the Lucchese venture, see C. Guasti, *Commissioni di Rinaldo degli Albizzi*, III, Florence, 1873, pp. 440–515.

style through a series of signed and dated pages from his hand.[8] He wrote the chapter containing the speech of Brunelleschi in a late period, probably in the 1440s or perhaps in the early 1450s. However, the meeting of the two men must have taken place long before that time. Its record contains a clear echo of an event of 1425, a close reference to an event of 1427, and only foreshadowings of events of 1430. It also contains peculiarities and errors of types often incurred when an author transcribes earlier notes.

The event of 1425 is the Brunelleschi-Acquettini debate, which took place in or shortly before that year and surely some considerable time after Brunelleschi's ship invention patented in 1421. Some of the coarse epithets of the debate are quoted literally. The event of 1427 is Taccola's completion of an *experientia* or official test, applied to a claimed invention of his own.[9] Brunelleschi's counsel to Taccola that disclosures of inventions should be limited in the extreme may be considered wise or foolish, but it was, in any event, most timely in the year of Taccola's test proceedings related to an invention. The event of 1430 is known as the Lucca disaster. Brunelleschi and the Florentines, in a war against Lucca and Siena, tried to take Lucca by "converting the city into an island." They so converted it, but only to their own discomfiture.[10] That Taccola does not mention the affair in his report of Brunelleschi's speech, where river-deflection works are in the center of interest, indicates that the siege of Lucca had not yet taken place.

The following is a close and complete translation of Taccola's text, wherein Filippo is quoted and matters of interest to both men are then discussed:

Pippo Brunelleschi of the great and mighty city of Florence, a singularly honored man, famous in several arts, gifted by God especially in architecture, a most learned inventor of devices in mechanics, was kind enough to speak to me in Siena, using these

[8] Details are shown in the forthcoming publications. There is valid evidence of writings by Taccola beginning in 1424 (E. Romagnoli, "Biografia cronologica de' bellartisti senesi," unpublished manuscript L. II. 4, vol. IV, ca. 1830, in Biblioteca Comunale, Siena). Authentic dates occur in Books I and II of *De ingeneis* (1427), Books III and IV (1433). There are sketches added in Lat. 197 with dated entries in 1438 and 1441, and copies of *De machinis* establish 1449 as the year of completion of the autograph. The dates are corroborated by Sienese documents and Imperial records. A final dated record is Taccola's Sienese tax declaration (1453), a part of which is published in facsimile by C. Pini and G. Milanesi, *La scrittura di artisti italiani*, Florence, 1876, Pl. 51; full transcript in G. Milanesi, *Documenti per la storia dell' arte senese*, II, Siena, 1854–1856, pp. 284 f. The handwriting specimens in Pini-Milanesi, *La scrittura*, include those of Brunelleschi, at Pl. 21, and of Vasari, at Pl. 188.

[9] Lat. 197, fo. 61r. The invention relates to Floating Caissons.
[10] Poggio Bracciolini, *Istoria fiorentina*, Florence, 1492, ch. 6; N. Machiavelli, *Istorie fiorentine*, 1520, ch. 5. For details see Guasti, *Commissioni*, III, pp. 440–515. The affair is clearly alluded to in other writings by Taccola: Lat. 197, fols. 107r, 113v–114r, 115v. A copy of *De machinis* in Paris, Bibliothèque Nationale, Lat. 7239, fo. 11v, includes one of the allusions missing in Lat. 28800: . . . *potius est infamia facienti . . . .*

words:[11] Do not share your inventions with many, share them only with few who understand and love the sciences. To disclose too much of one's inventions and achievements is one and the same thing as to give up the fruit of one's ingenuity. Many are ready, when listening to the inventor, to belittle and deny his achievements, so that he will no longer be heard in honorable places, but after some months or a year they use the inventor's words, in speech or writing or design. They boldly call themselves the inventors of the things that they first condemned, and attribute the glory of another to themselves.[12] There is also the great big ingenious fellow, who, having heard of some innovation or invention never known before, will find the inventor and his idea most surprising and ridiculous. He tells him: Go away, do me the favor and say no such things any more—you will be esteemed *a beast*.[13] Therefore the gifts given to us by God must not be relinquished to those who speak ill of them and who are moved by envy or ignorance. We must do that which wise men esteem to be the wisdom of the strong and ingenious:

We must not show to all and sundry the secrets of the waters flowing in ocean and river, or the devices that work on these waters. Let there be convened a council of experts and masters in mechanical art to deliberate what is needed to compose and construct these works. Every person wishes to know of the proposals, the learned and the ignorant; the learned understands the work proposed—he understands at least something, partly or fully—but the ignorant and inexperienced understand nothing, not even when things are explained to them. Their ignorance moves them promptly to anger; they remain in their ignorance because they want to show themselves learned, which they are not, and they move the other ignorant crowd to insistence on its own poor ways and to scorn for those who know. Therefore the *blockheads* and ignorants are a great danger for the aqueducts, the means for forcing the waters, their ascending and descending both subterranean and terrestrial, and the building in water and over the water, be it salt or fresh. Those who know these things are much to be loved, but those who do not are even more to be avoided, and the *headstrong* ignorant should be sent to war. Only the wise should form a council, since they are the honor and glory of the republic. Amen.

Note[14] that there are two things wherein the river itself causes difficulty when you want to build the palisade for a mill [pond] or to construct bridges. They are these:

[11] The statement following here may be the only utterance orally made by Brunelleschi that was literally transmitted by a direct witness. Of course, Taccola translates it into his notarial Latin. Words in italics indicate Taccola's use of Italian within his Latin text.
[12] An apparent reference to the debates with Lorenzo Ghiberti.
[13] Brunelleschi seems to paraphrase Acquettini's sonnet.

[14] From here on to the end of the folio, the entire text is in Italian.

Firstly, the [need to] deflect the river in order that it may not carry away the work you do; secondly, [build] such construction of the riverside that the water may not eat away the abutments of the bridge or the wall and foundation of the palisade masonry. Thirdly, if you cannot deflect the river from its bed, leave part of the palisade open and unfinished, to let the river pass through it, so that it may not interfere with your construction. Arrange it to confine the river by a fall door or sluice gate. In the fourth place, see whether the river bed has large pieces of tufa in it. If material of that kind is in the river, it will be better not to drive piles, because the piles would break the bed of such a river. In that case, when the structure has been built and when high water comes, the water would find the area broken by the piles. The flow then carries everything away, the structure remains on the piles, and water runs under the foundations. Now as to foundations with underpinnings in a river bed. If the river has soft ground, or muddy or sandy places, or salty or brackish water, timbers are [to be] found, since such places [call for] piles driven into the ground, whereon the bridge abutments or palisade walls can be founded. Another thing: when the wall is freshly made, remember to cover it with boards, in order that the river water may not carry away your work. When first preparing such a site, build a shed to protect the workshop and the masters' tools, and [to serve as a place] where they can stay, etc. Also when building the riverworks cited

above, look first for forests nearby, where there are places for timber, and see whether there is stone for walls and for making lime, as well as gravel or sand for the masonry. If these things are not close by, he who contracts for such a construction or who undertakes it will become poor if he had been rich. I hold him to be wise who will not undertake it [to build riverworks] when no estimate can be made of the cost of all timber, stone, lime, furnaces, pack-hauling, transportation, cartage, manufacture, habitation, and sustenance. If you take it on, arrange everything with those who will do the work, etc.

At the bottom of the page on which the speech begins is a drawing of a pressure-operated water duct. It illustrates the topic of Brunelleschi's remark on "aqueducts" and their "ascending and descending," which follows on the next page. This form of water duct appears at other points in the Taccola books and in later books of Francesco di Giorgio and others.[15] It is a schematic illustration of the famous *butini*, or underground aqueducts of Siena.[16] One part of Brunelleschi's speech, as quoted by Taccola, ends with "Amen" on the page where

[15] Taccola: Lat. 197, fols. 113^v–114^r, 115^v, 116^v; Palat. 766, 48^v–49^r; Lat. 28800, fols. 35^v–36^r. Francesco di Giorgio: Urb. Lat. 1757, fols. 8^r, etc. Antonio Averlino, known as Filarete, *Treatise on Architecture*, II, ed. J. R. Spencer, New Haven, 1965, fols. 94 f.

[16] F. Bargagli-Petrucci, *Le fonti di Siena e i loro acquedotti*, Florence, 1906, *passim*. The manuscript, Lat. 197, fo. 107^v, also relates to waterworks of Toledo, as do certain copies based on Taccola's book.

discussion of aqueducts begins, and the next part concerning river works is probably its continuation. It seems that the two men exchanged experiences about the building of waterworks, Taccola contributing his or his countrymen's knowledge about the *butini* and Brunelleschi giving his own views about such matters. There is no other good explanation for Taccola's turning away from Latin to an Italian writing at the start of these hydraulic technicalities. The use of this idiom is most exceptional, as he invariably uses Latin, while Brunelleschi, according to his biographers, always used the vernacular.

There are small sketches on the Brunelleschi-*butini* page, the nature of which is not clear. Conceivably they belong to the Brunelleschian discussion of hydraulic constructions. Still other sketches appear on adjacent pages and are probably connected with Brunelleschi's work. Several of them show the deflection of a river into a town, the unsuccessful stratagem used at Lucca.

Taccola also discussed stratagems at many points in his text, frequently deriving them from classic works or their medieval restatements.[17] There is some evidence that he discussed this area, too, with Brunelleschi and derived from him the suggestion for one of the most celebrated pictures in his book. He shows a fortress exploded by gunpowder in an underground mine, carefully showing the powder kegs and fuse, and clearly describing each part of the procedure. This proposal for an explosion goes back to Domenico di

[17] See Berthelot, "Pour l'histoire," *passim.*

Matteo, an acquaintance of Brunelleschi.[18]

The principal invention of a peaceful kind, discussed between Taccola and Brunelleschi, is the construction of buildings "in and above water." It is illustrated in Taccola's schematic style on pages directly facing both parts of Brunelleschi's speech (Figure 38, right parts) and is obviously connected with the speech, although Taccola does not state whether he still quotes the master. He writes,

Foundations of walls constructed in water, etc. When you build the foundation of a wall on water, take trough-shaped beams of oak and place them on the water, putting strong boards thereon.[19] Put barrels on the boards, and then place the masonry of stones, cemented with lime and well-washed river sand. Make such a wall as long as the interlocked woodwork, and up to eight feet high by five feet thick, and no more; otherwise there is too much weight on the interlocked beams and they sink to the bottom of the sea or river, etc.[20]

The challenging expression used here by Taccola, "when you build the foundation of a wall on water," is introduced without explanation; but it

[18] Gaye, *Carteggio,* I, pp. 86 f., 126. Taccola's illustrations are in Palat. 766, fols. 60r, 61r, and Lat. 28800, fols. 33v, 34v. The views of modern historians about Taccola's development of this device, and the work of his early successors, are full of erroneous speculations.
[19] The text no doubt means nailing the boards across the beams.
[20] The "etc."—which Taccola uses frequently—seems to imply that addenda are to be found elsewhere in Taccola's work.

is obviously related to that part of Brunelleschi's speech where he refers to "building in water and over the water." Did Brunelleschi, or Taccola, build such structures?

Considerable research will be needed to determine whether the terms used here are connected with Brunelleschian work in the then-current construction of towers in Porto Pisano, connected with the Medicean developments in maritime work that we noted in another study. Part of the work is sometimes attributed to Brunelleschi, or to Ghiberti.[21]

Taccola's interest in such work is ubiquitous in his treatises. For example, he adds a further illustration directly following the present ones (fo. 110r).[22] Below it he gives additional instructions, including the point that the wall should have "barrels in its masonry, with air enclosed therein, to allow the wall to float." Other disclosures make it clear that the wall, constructed on a floating support, is ultimately submerged, and that various alternate procedures can be used in suitable cases.[23] If a wall is built on a floating caisson, this is preferably done near the shore, where materials are readily at hand. Thereafter, caisson and wall are floated to the proposed site. Valves are then opened in the caisson to cause the structure to sink to the ground, which has been dredged and flattened beforehand. The process is known today as floating-caisson construction. It may be traced to earlier prototypes, both medieval and ancient.[24] Taccola, in 1427, tested applications or improvements of this process, which he claimed as his invention. It appears now that he discussed his invention with Filippo about the time of the test.

Further improvement is then suggested when the text speaks of "interlocked beams" for the floating caisson, and of "barrels" in the masonry. Conceivably these are ideas arising from Brunelleschi's experience with tie rings for a hollow-wall structure. This type of structure then became a starting point for further developments.[25] Similarly there was much interest for the alternate processes, including the one employing a ballast-loaded "palisade"

[21] Some pictures and facts about Porto Pisano are supplied by Guarnieri, *Da Porto Pisano*, pp. 34 f., 41, 56, 61 ff., 102 ff., and opposite p. 160.
[22] On fo. 109¹ he once refers to "fo. 228." This was his original pagination number, later replaced by 107.
[23] Palat. 766, fols. 42r, 42v, 43r, 43v.

[24] *Mappae clavicula*, ed. T. Philipps, *Archaeologia*, XXXII, 1847, pp. 183 ff., 209; Heron of Alexandria, *Mechanica*, III, 11. It seems remarkable that the author of *Mappae clavicula*, or Taccola, or Brunelleschi, knows the method described in Heron's book, supposedly surviving only in Arabic copies, but is totally silent about all parallel proposals of Vitruvius. The problem of rediscoveries of these technical books during the Renaissance is in need of searching review. It becomes yet more complex because the *Mappae clavicula* is bound with an eleventh-century Vitruvius in Sélestat, and the Vitruvian part was probably known to the Ghiberti's. Also see D. L. Hoffman, "Pioneer Caisson Building," *Journal of Society of Architectural Historians*, XXV, 1966, pp. 66–71.
[25] Strauch, *History of Civil Engineering*, pp. 19 f., 83 f., 132; R. S. Kirby and P. G. Laurson, *The Early Years of Modern Civil Engineering*, New Haven, 1932, p. 135; D. B. Steinman, *Bridges*, New York, 1957, pp. 88, 101.

structure.[26] All this was discussed with Filippo and so was the "rising and falling of water under ground." This final concept is based on scholastic speculations, which were considered "science." The idea was based on legends and the wisdom of mystics. In substance, Taccola's formulation and drawing are an abbreviated and simplified version of material from a book that a minor scholastic author produced in the eleventh century. Taccola writes,

Defense of Christianity. When it is attacked by the infidels, aid must be had from Prester John in India. He must be asked to seal off the rivers that pass through his gates to the territory of the Tartars, and likewise the gates that discharge the rivers running to the Sultan's territory, so that their territories remain dry. If the Sultan asks Prester John on some ground to leave the gates open, refer the Sultan directly (*ipse*) to the war against the Turk, etc.

It is my view that in the center of the earth there is natural fire, which is the soul of the entire earth, whereby the entire earth is divided as I said before when speaking of the universe. In this [central] place all elements have their origin, such as metals and sulfur, wherefrom hot waters have their origin. Fire always burns there, as can be seen by faithful witness accounts from Moncibello,[27] near the city of Catania, where fire issues from the volcano with enormous roar. Earth

is lifted up by the air enclosed in its cavities and pores. Flames of fire and air naturally rise to the upper regions, etc.[28]

Part of this curious statement reappears in a later treatise of Taccola, and in its copies, without the passage about "the Turk." Strangely, one of the copies later came into Turkish hands.[29]

Not all the "scientific" traditions known to Brunelleschi and his followers were of such mystic kind as the compositions about Prester John, his mighty rivers, and the central fire. These traditions also included laws or assumed laws of physics, for example, those about hydraulic siphons, which Taccola tried to demonstrate experimentally and to express quantitatively.[30] To modern readers it may seem that only the more constructive ones of these assumed laws were a real beginning or renewal of science. However, all the traditions inherited from the past were fascinating to early explorers and were conducive to some of their explorations. All of them contributed to discoveries

26 Palat. 766, fols. 42r, 42v; also Lat. 197, fo. 109v.
27 Another name for Etna, from *mons* and *gebel*, both meaning "mountain."

28 Although the second paragraph is ultimately derived from Empedocles, through Aristotle, its present form is copied literally, with abbreviation, from the *Imago mundi* of Honorius Solitarius, ca. 1090. (About Honorius, see G. Sarton, *Introduction to the History of Science*, I, Baltimore, 1927, p. 749.) A transcript of the *Imago mundi* is in a copy of Taccola's *De machinis*, Lat. 7239, which also contains another version of this present paragraph about the universe. Perhaps Taccola owned a copy of the *Imago*; perhaps he talked about it with Brunelleschi.
29 For the basic history of the copy, Lat. 7239, now in Paris, see Thorndike, "Marianus," pp. 7 ff.
30 Palat. 766, fols. 32r, 57r; Lat. 197, fols. 68r, 73v, 75r, 115r.

and inventions that gradually eliminated the inherited errors, both big and
small. Thereafter men could ask more
significant questions, as did Galileo, and
in due course the simple and elegant
answers were found that became part
of modern science.[31] Enormous efforts
were needed to transform the quasi-
science of the dark ages into the science
of modern times. A significant part of
these efforts, although an early and
remote one, may be identified with
Brunelleschi's work and that of his
friends or associates.

[31] We are referring to the scholarly discoveries of Federigo Commandino, the inventions of Giovanbattista della Porta, and
the famous barometric demonstrations of
Torricelli and Pascal. See, for example,
C. de Waard, *L'expérience barométrique*,
1936.

CONCLUSIONS

When the Florentines buried Filippo in the Cathedral that he had substantially completed for them, they asked his pupil and adopted son Andrea il Buggiano to show his likeness in marble (see the frontispiece of this book) and below it they wrote,

> How Filippo the architect excelled in invention is shown not only by the beautiful shell of this famous temple but also by various machines that he invented with divine genius. . . .

Ever since then he was known as a most remarkable engineer and as the man who worked most outstandingly in one architectural style while laying the foundations of another. In our opinion he was one of the great developers of building principles for all times.

Before him there had been an impasse in Florentine architecture. Romanesque and Gothic builders had proposed and debated ambitious plans, but only the Gothic masters had specified clear ideas about a possible way of execution of the plans, and one basic part of their formula, the use of outer buttresses, had been vetoed by the Romanesque masters.

A full return to Gothic forms and methods, for the construction of supports of the Cupola, was proposed by Giovanni d'Ambrogio, who directed the work during the first eighteen years of the Quattrocento. Brunelleschi's Cupola structure is different from that suggested by Giovanni, but even in Brunelleschi's work there is considerable Gothic influence. Far from using a single tradition, either in statics or ornamentation, his Cupola uses a structure and form wherein Gothic and Classic elements are synthesized. This fact is often overlooked. It is misleading to call Filippo merely the renewer of an ancient form.[1]

The history of the building demonstrates the recurrence of medieval and ancient traditions and the master's achievement in uniting them. In 1404, the tall buttresses and windows of Giovanni d'Ambrogio were rejected, but the lower buttress elements shown by the model of his predecessors were retained (as shown in our Figure 5). This was done by a group that included Brunelleschi and Ghiberti. A few years later the Florentine Tambour was designed, approved, and built by new groups of men, whose identity is less clearly documented. According to eminent writers, Brunelleschi was influential (again). We think there is good evidence in support of this view, and there is no good evidence against it. We may conclude that Filippo, Lorenzo, and perhaps others, jointly opposing Giovanni d'Ambrogio, established a wide spatial separation between the compromise features used in the understructure and a new, higher,

---

[1] As mentioned by *Sanpaolesi*, pp. 110 f., Filippo brought about a repudiation of Gothic art. However, as noted by *Sanpaolesi*, p. 31, Filippo himself was interested in Gothic-influenced statics. A similar view is stated, although unclearly, by P. Frankl, *Die Renaissance-Architektur in Italien*, Leipzig, 1912, p. 2. In fact, as indicated in these studies, Filippo contributed creatively to traditions in statics which are of a peculiarly Gothic character.

independent Cupola. Like the artists of 1367, these men vetoed a conventional Gothic form, without showing exactly what form was to be substituted. It was not necessarily a form totally devoid of Gothic features.

Soon after completion of the Tambour and of some model work for statues to be placed in lower regions of the Dome, or perhaps after abandonment of work on these statues, Filippo showed the form to be used for the Cupola and began to persuade the builders to adopt it—a process that lasted from 1417 to 1420. He proposed to construct the large vault as a double shell reinforced by ribs and small arches; these were modified Gothic buttresses and hollow ribs. He combined them with tie rings; these were modified Roman features. He fully overcame the former hesitations of the Florentines and persuaded them to adopt the new design, which he then used with singular success in building the Cupola during the next twelve years. He also began to produce final details, including the Classic-Gothic design for the Lantern crowning the Cupola. His structure solved the static problems that he had inherited, and foreshadowed basic modern principles of construction. It gave timeless expression to a new architectural principle—a major addition to the other architectural innovations introduced by Filippo. At the outset of these studies we were not sure whether Brunelleschi's achievement was as great as his friends asserted. We are now inclined to think that he may have been more creative as a builder than most of his admirers imagined.

Filippo also devised mechanical systems for delivering, hoisting, and placing the building materials. Some of these devices may have contributed to the progress of machinery and surely they contributed much to his personal fame, as they allowed him to build with remarkable speed and safety. His mechanical inventions also had side effects in the evolution of patents. The machines were admired, and their success gave confidence to other mechanics.

We have tried to develop an image of Brunelleschi the builder and engineer, but had to disregard many facets of his remarkable personality, such as his contacts with goldsmiths, sculptors, and painters. We are able to cite only two personal, verbal statements of Brunelleschi the man—a sonnet deriding a critic and a speech addressed to a friend. We see him use playful superiority over his critic and hear him talk to his friend about the people at large with a strange mixture of sarcasm and rage. The biographers add, credibly, that he spoke freely with common people and, for a purpose, with men of property and influence; that he hated his rival; and that he made his points with single-mindedness and insistence. In his specification for the Cupola he makes his points in nearly perfect form. Our impression is that his powers came from what he saw rather than from that which others could tell him.

Brunelleschi's work, together with the work of direct followers, almost totally transformed the artistic traditions of Quattrocento builders. His influence on building technique may have been comparable in depth, if not in rapidity. However, much remains to be studied in this technical area.[2]

The engineering work of the Renaissance began to be studied near the end of the eighteenth century, when G. B. Venturi rediscovered the "mathematical-physical" works of Leonardo, together with those of Taccola. Since then many documents relating to such works have been found. The hero worship and anecdotal accounts of early biographers then fell into some disrepute. However, we must not be too harsh with the biographers. The rediscovered record proves that interesting beginnings were disregarded by them but not that Filippo's life and work was significantly different from the pattern established by the early reports. The biographers and their informants among the common people did not use words of precise definition— nor did the early architects and notaries do so when they produced the documents, rediscovered in recent times. Only when official documents, engineering notebooks, or biographies independently corroborate one another do these records gain more than the limited evidentiary value that each of them inherently has.

The studies undertaken here, using biographies, drawing records, and official documents, have led us to the conclusion that many historic processes came under Brunelleschi's influence. We hope that future historians will consider his time and work in a comprehensive way. We find the effort rewarding, even if the conclusions remain tentative. When we first undertook our studies, we were ignorant of several Brunelleschian reports that now are frequently mentioned, and many such reports are still imperfectly known. The notes of Taccola, Ghiberti, and Leonardo are among them, aside from those of Alberti, Francesco di Giorgio, the San Gallos, and many others. The study of some such reports alerted us to interesting facts and challenging problems. Even so we are not sure that we have found more than a sampling of what can be found on further research. Records of several centuries remain to be reread and reevaluated. Their study should help us, gradually, understand more of the growth of architecture and technical science brought about by Brunelleschi and his contemporaries.

[2] See the references to civil and mechanical engineering in the chart presented at the end of this book. It outlines pathways of Brunelleschian influence, as they appear to us from these studies. Each name listed in the chart stands for many concepts proposed or elaborated or transmitted; few of the details have been studied thus far.

DOCUMENTS

I. SPECIFICATION FOR THE CUPOLA, 1420[1]

. . . Consules una cum operariis . . . in palatio dicte Artis collegialiter congregati . . . considerantes modellum de quo infra fit mentio, credentes supra forma et tenore ipsius modelli suo recto ordine procedere . . . habitoque . . . conloquio et consilio . . . providerunt ordinaverunt et deliberaverunt,

Quod per dictos offitiales ad constructionem cupule antedicte procedatur . . . quemadmodum continetur et fit mentio in modello infrascripto. . . . Cuius quidam modelli volgare sermone scripti tenor talis est:

Qui appresso fareno memoria particularmente di tutte le parti si contengono in questo modello facto per esempro della cupola magiore.

Imprima la cupola da lato dentro e volta a misura del quinto acuto neglangoli. Ed e grossa nella mossa da pie braccia tre e quarti tre. E piramidalmente segue siche nella fine congiunta nell'occhio di sopra rimane grossa br. 2½.

Fassi una altra cupola di fuori sopra questa per conservalla dal umido, e perche torni piu magnifica e gonfiante. Ed e grossa nella sua mossa da pie braccia uno e quarti uno; e piramidal-

mente segue in sino al'occhio di sopra rimane braccia ⅔.

Il vano che rimane tra l'una cupola e l'altra si e dappie br. 2 nel quale vano si mectono le scale per potere cerchare tutto tra l'una cupola e l'altra, et finisce il decto vano al'occhio di sopra braccia 2⅓.

Sono facti 24 sproni, cioe 8 neglangoli e 16 nelle faccie, ciascuno sprone deglangoli e grosso da pie braccia 7 dalla parte di fuori, e nel mezzo di decti angoli in ecascuna faccia si e due sproni, ciascuno grosso dappie braccia 4, e legano insieme le decte due volte, e piramidalmente murate insino alla somità dell'occhio per iguale proporzione.

I decti ventiquattro sproni colle decte cupule sono cinti intorno di sei cerchi di forti macigni e lunghi e bene sprangati di ferro stagnato, e di sopra a decti macigni sono catene di ferro che cerchiano intorno le decte volte con loro sproni. Assi a murare di sodo nel principio braccia 5¼ per alteza e poi seguire li sproni.

Il primo e secondo cerchio e alto braccia 2, el terzo e'l quarto cerchio si e alto braccia 1⅓, el quinto e 'l sexto cerchio alto braccia 1; mal primo cerchio dappiè si e oltraccio aforzato con macigni lunghi per lo traverso, siche l'una cupola et l'altra si posi in su decti macigni.

E al alteza dogni dodici braccia o circa delle decte volte sono volticciuole a botti tra l'uno sprone e l'altro per andito intorno alle decte cupole e sotto le dette volticciuole tra l'uno sprone e l'altro sono catene di quercia grosse

[1] The text following here is from Arte di Lana, Partiti, atti e sentenze, vol. 149, fols. 59ᵛ, 60ʳ, Archivio di Stato, Florence, discovered by Doren in 1898. We compared it with a copy of the text in Magl. XIII, 72, fols. 37ᵛ, 38ʳ, Florence, Biblioteca Nazionale, which differs only in minor details from this "Partiti" text. See the translation, pp. 32 ff.

che legano i decti sproni e in su decti legni una catena di ferro.

Gli sproni sono murati tucti di macignio e pietra forte, e mantegli overo le faccie delle cupole tutte di pietra forte legate cogli sproni per insino al'alteza de braccia 24, e da indi in su si murera di mactoni o di spugna, secondo si deliberra per chi allori l'ara a fare, ma piu legiere materia che pietra.

Farassi uno andito di fuori sopra gli otto occhi di sotto imbecchatellato con parapecti trasforati, e d'altezza di braccia 2 o circa al'avenante delle trebunecte di sotto; o veramente due anditi, l'uno sopra l'altro, in su una cornice bene ornata, e l'andito di sopra sia scoperto.

L'acque della cupola termino in su una racta di marmo, larga uno terzo di braccio e gitti l'acqua in certe doccie di pietra forte murate sotto la racta.

Farannosi 8 cresste di marmo sopra glangoli nella superficie della cupola di fuori, grosse come si richiede e alte braccia 1 sopra la cupola, scorniciate e a tecto, larghe braccia 2 di sopra, sicchè braccia 1 sia dal colmo alla gronda d'ogni parte, e murisi piramidali dalla mossa insino alla fine.

Murinsi le cupole nel modo sopradecto sanza alcuna armadura, massimamente insino a braccia trenta, ma con ponti, in quel modo sarà consigliato e diliberato per quegli maestri che l'aranno a murare; e da braccia trenta in su secondo sarà allora consigliato, perchè nel murare la praticha insegnerà quelle che ss'ara a seguire.

II. SPECIFICATION FOR THE CUPOLA, 1421–1422[2]

. . . Consules . . . una cum operariis . . . et quattuor officialibus . . . providerunt . . . infrascripta in vulgari sermone scripta, vid.:

Che sedici sproni, cioe due per ciascuna faccia della cupola, dove altra volta si dilibero che fossero di grosseza di braccia quattro l'uno da lato di fuori, per levar via carico superfluo si faccino di braccia tre l'uno da la parte di fuori.

E che le cupole, dove altra volta si dilibero si facessono di pietre per insino a l'alteza di braccia XXIIII, per levar via el troppo carico e peso, si faccino di pietra per insino coperto sopra i cardinali degli usciuoli s'anno a murare al presente, che sara circa di braccia dodici alto della mossa delle cupole: e da indi in su si murino di quadroni, cioe mattoni.

III. SPECIFICATION FOR THE CUPOLA, 1425–1426[3]

. . . Consules . . . atque . . . operarii . . . ac etiam offitiales dicte cupole . . . visis . . . consiliis et modellis factis . . . et visis quibusdam scriptis[4] . . . et viso

[2] Book BD 81, fo. 17ᵛ of the Opera, copied in *Cupola*, Doc. 52. Dated 13 March 1421-1422. See a partial translation; pp. 33, 35
[3] Book LD I, fols. 170ᵛ, 171ʳ of the Opera copied in *Cupola*, Doc. 75. Dated 24 January 1425-1426 (Report) and 4 February 1425-1426 (Decree). Also see pp. 43 ff.
[4] Among the "counsels, models and writings," preceding the Report of 24 January 1425-1426, was the memorandum of Acquettini wherein he asked for construction

quodam raporto . . . cuius tenor talis est:

MDCCCCXXV adi XXIV di gennaio

Raporto facto a voi sig. operai e ufficiali della cupola . . . per Filippo di Ser Brunellesco, Lorenzo di Bartaluccio e Battista d'Antonio capomaestro insieme d'accordo con Giuliano di Tommaso Ghuccio sopra la commissione a loro data per voi etc.[5]

In prima che in sul secondo andito della cupola maggiore, dove al presente è fatto la catena de'macigni, in ongni faccia di detta cupola si facci uno occhio di diamitro d'uno braccio, per comodo di fare ponti al musaico s'à a fare, o d'altro lavorio[6] . . . . Anchora che sopra i cardinali degli usciuoli che sono sopra 'l detto secondo andito, per perfectione del cerchio che gira intorno la cupola di fuori, accio che detto archo vivo sia intero e non rotto, si muri di mattoni in atto d'archo, di grosseza quanto e la detta cupola di fuori e su a alto braccia uno circha (Et se mai paresse che detta aggiunta mostrasse rustica a l'occhio, o impedisse l'andito e schale, si possa, fatto la cupola, disfare detta aggiunta) accio che con piu sicurta si possa

guidare a murare la cupola insino alla fine.[7] Anchora in ogni faccia della cupola si muri due catene di macigno di largheza e alteza tre quarti di braccio, o meno, che contengha di lungheza quanto e l'una cupola e l'altra, cioe e sopra due sproni che vanno nelle faccie. Et sopra dette catene di macigno si pongha una catena di ferro per ciaschuna, che contengha la lungheza de' macigni.[8]

---

[7] "[At a level] above the lintels of the doors on said second walkway, for the perfection of the tie ring around the outer cupola and to insure that its effective curve may be an integral and uninterrupted whole [in spite of said doors], masonry shall be placed in arc-shape, as thick as said outer cupola and rising to a height of circa 2 feet, whereby the Cupola masonry may be placed and completed with greater safety. (The addition can be removed on completion of the Cupola if the addition be found to appear ugly or to interfere with the walkway and stairs.)" Guasti writes: ". . . a alto braccia uno circha. E se mai paresse che detta agiunta . . . impedisse l'andito e schale, si possa . . . disfare detta agiunta, acioche . . ." This would be meaningless. In our opinion the phrase "et se . . . disfare detta agiunta" is parenthetical. For greater clarity, we have placed it after the phrase "whereby . . ." in our translation. It is conceivable that the axes of the "arc-shapes" ran radially across the shell and that the additions were blind arches projecting outwardly and inwardly until the projections were removed. However, it seems more probable that the axes were parallel to the ribs and the additions are the *volticciuole* spaced from the previous ones by 13 *braccia* instead of 11 *braccia*.
[8] We mentioned these stone beams in a note relating to the original plan (specification of 1420, point 6). It is not clear from the present document whether the stone beams run radially across the shells or in the shells, that is, whether we have here a substantial addition to the original

---

of twenty-four windows in the Cupola (see our discussion of the stone chains in the masonry of the Cupola). This document mentions Acquettini, along with many others, as an author of "counsels, models and writings."
[5] Giuliano di Tommaso Ghuccio acts as notary.
[6] The text adds: . . . *per molti altri cittadini.* A note in *Cupola*, p. 39, suggests that *comodi* may be meant.

Anchora si facci fare mattoni grandi
di peso di libre venticinque insino a
trenta l'uno e non di piu peso, i quali
si murino con quello spinapescie sara
diliberato per chi l'ara a conducere. E
da lato della volta dentro si pongha
per parapetto assi che tenghino la
veduta a maestri, per piu loro sicurta.
E murisi con gualandrino con tre
corde, faccia dentro e si di fuori. Non
si dice alcuna cosa de' lumi perche
s'imagina vi sara lume assai per gli
otto occhi di sotto, ma se pure nel
fine si vedesse bisognasse piu lume, si
puo argomentarlo[9] dalla parte di sopra
agievolmente a lato a la lanterna. Ne
si dice anchora di farla cientinare. non
che non fosse suto [?] piu forteza de
lavorio, e piu bella, ma non sendo
principiato, parebbe ch'il centinare al
presente, lavorio straordinario da
quello ch'e murato e mostrerebbe
altra forma, e anche difficilmente si
potrebbe centinare sanza armadura,
perche 'l centinare si lascio di principio solo per non fare armadura ecc.
E se presto delle precedette cose si
piglia partito si puo seguire il lavorio
a marzo.

Io Giuliano di Tommaso di Ghuccio
sopra-detto scripsi le sopradette cose
di volere de' soprascripti, dì detto.
. . . decreverunt quod laborerium
. . . sequatur . . . . ac etiam, non
obstante prefato raporto . . . con-
cesserunt . . . potestatem . . . in

addendo, minuendo ac disponendo . . .
prefatum laborerium . . . .

IV. BILL OF MATERIALS FOR FILIPPO'S
HOIST, 1421[10]

Filippo di ser Brunelescho dè avere per
spese fatte nello edificio da tirare,
chome partitamente apresso diremo:

A di 26 d'aghosto, lire 3 a Montino di
Bruogio, per trainare 1° olmo per lo
subbio del chanapo.

A di 30, soldi 20 portò il Testa scharpelatore, per ghabella di 2 ruote.

A di detto, lire 6 portò Papi di Sandro
scharpelatore, per braccia 6 di quercie
per fare le chasse de' bilichi.

A di detto, soldi 16 ebe Ghuido da
Norcia pontatore,[11] per rechare le
ruote di Verzaia.

A di detto, soldi 5, denari 6, portò Papi
di Sandro, per fare rechare braccia 6
di quercie.

A di 2 di settenbre, soldi trentatre a
Marino di Benedetto legniaiuolo per
manifatura di 4 chasse pe' bilichi.

A di 11 di settenbre, ebe 1° portatore,
per rechatura di 3 chavigli per la ruota,
soldi 4.

A di 17 di settenbre, lire 2 ebe Manno
di Beninchasa legniaiuolo per 2 girelle
di nocie, dove si posa lo stile del
chanapo.

A di 24 di settenbre, lire 13, soldi 10
ebe Tano di Bartolo legniaiuolo per 3
chastagni che sostenghino l'edificio.

A di deto, soldi 6 ebe Fede charetiere,
per rechare i deti chastagni.

A di 19 deto, soldi 8 per ghabella della
vite.

A di 29, soldi 20, per ghabella del subio
cholle ruote.

A di 9 di novembre, soldi 12 ebe

---

point 6 or a dimensional change to this
point. *Sanpaolesi*, p. 55, at C, seems to
interpret it as an addition. He does not,
however, show the more basic stone-chain
structure, and does not describe it.
[9] *augmentarlo* ?

[10] Book BD LXXX, fo. 71[r] (Decree) and
Book BS, BB, fols. 118[r], 118[v] (Bill of
Materials), copied in *Cupola*, Doc. 125.
Partial translation: p. 87.
[11] *portatore* ?

Andrea di Franciescho fabro, per 1°
pezo di chatena.

A di 13 di novembre, lire 11 ebe
Nanni di Franciescho legniaiuolo per
1° olmo per lo stile del chanapo.

A di detto, soldi 16, denari 6, per
ghabella di detto olmo.

A di detto, soldi 50 per fasciare 2
gioghetti di chuoio, e per 1° soiatto per
lo bue, ebe Chante sellaio.

A di 23 novenbre, lire 3, soldi 16, ebe
Piero di Ciuto seghatore e 'l conpagno,
per due opere per seghare legname per
lo deto edificio.

A di detto, lire 3 p. ebe Buono di ser
Bencivenni, per parte di cieste da
cholla.

A di 26 di novembre, lire 2, soldi—,
denari 8, ebe Jacopo d'Andrea leg-
naiuolo, per braccia 12 di chorenti di
fagio, e braccia 15 1/1 d'asse di faggio.

A di 20 di dicienbre, soldi 35 ebe
Lorenzo di Nicholo legniaiuolo, per
due ghobi da nave per lo timone de'
buoi.

E a di 9 di giennaio, soldi 5, per fare
seghare 1° olmo.

A di 7 di febraio, soldi 29, per 4 churri
per ghuardia del chanapo, conperati
da Lionardo di Giovanni torniaio.

A di 10 di febraio, soldi 7, per fune,
conperò da Matteo schodellaio.

A di 11 di febraio, soldi 15, ebe Maxo
di Chincho, per 1° gioghetto per lo
secondo bue.

A di 15 di febraio, lire 3, soldi 10, ebe
Chanto di Chavalcanto, per 1° soiatto
grande per lo bue.

A di 10 di marzo, lire 40, soldi 12, ebe
per 14 bighoncie da cholla, conperò da
Fruosino d'Andrea bottaio, a soldi 58
l'una.

A di detto, per 2 bilichi per le ruote a
charuchole, libre 70 1/1 a soldi 4 libra,
tolse da Mattio fabro; montano lire 14
soldi 2. Abati denari 4 per lira, resta
lire 13, soldi 17, denari 8.

E dè avere fiorini 11, lire 56, per piu
legniame conperò da Maxo di Chirico
fa i charri, apartenente al detto edificio.

E dè avere lire 5, soldi 2, ebe maestro

Antonio Stoppa, per manifattura della
vite de lo edificio.

E dè avere lire 30, soldi 12, ebe
Antonio di Tuccio torniaio, per
manifattura di 91 palei a soldi 4 l'uno,
e per 16 mozetti a soldi 8 l'uno, e per
legname lire 6.

E dè avere lire 151, soldi 1, denari 8,
per opere 67⅓ di maestro Piero de'
Bianchi, a soldi 20 l'una; e per opere
67 di maestro Antonio de' Bianchi, a
soldi 25 l'una; le quali opere misono a
fare lo deto edificio.

E dè avere per libbre 1022 di feramenti
di piu ragioni, tolti da Giovan di
Fruosino fabro, per soldi 4, denari 4
libra, e per manifattura di 2 bilichi per
la ruota ritta, libre 78 a soldi 2 libra;
monta in tutto lire 229, soldi 4, denari
8. Abatesi denari 4 per lira, resta lire
225, soldi 8, denari 4.

Soma, fiorini 11, lire 584, soldi 12,
denari 5.

Stanziati a di 20 d'aghosto, per mano di
ser Dino di Chola notaio de l'Opera.

V. SONNET BY GIOVANNI DI GHERARDO, CALLED
ACQUETTINI[12]

O fonte fonda et nissa d'ignioranza,
  pauper animale et insensibile,
  che vuoi lo 'ncierto altrui mostrar visibile,
  ma tua archimia nichil abet substanzia.
La insipida plebe, sua speranza
  ormai perduta, la 'ede credibile,
  ragione non da, che lla cosa inpossibile
  possibil facci huomo sine substanza.
Ma se 'l tuo badalone, che 'n acque vola,
  viene a perfezione—che non può essere,—
  none ched' i' lega Dante nella schuola,[13]
Ma vo' con le mie mani finire mio essere:
  perch' io son cierto di tuo mente fola,
  che poco sai ordire e vie men tessere.[14]

[12] From Guasti, Belle arti. Translation: p. 118.

[13] There seems to be faulty copying at this
point of the manuscript, as also in the first
line where the meaningless nissa appears.
At present, we interpret non chied' io leg-
[g]er[e].

[14] Literally translated, this probably means,
"Little do you know about preparing a
frame [of a loom or other machine] and
even less about weaving."

VI. SONNET BY FILIPPO DI SER BRUNELLESCO[15]

Quando dall' alto ci è dato speranza,
  o tu c' hai efigia d'animale resibile,
  perviensi all'uomo lasciando il corruttibile,
  e ha da giudicare somma possanza.
Falso giudicio perde la baldanza,
  ché sperienza gli si fa terribile:[16]
  l'uomo saggio non ha nulla d'invisibile
  si non quel che non è, perch' ha mancanza
Di quelle fantasie d'un non si di scuola.[17]
  ogni falso pensier non vede l'essere
  che l'arte dà, quando natura invola.[18]
Adunque e' versi tuoi conviene stessere,
  che non rughiano il falso alla charola,
  dopo che 'l tuo inpossibile viene all'essere.

VII. RULES FOR INVENTORS AND BUILDERS[19]

(Fo. 107ᵛ) Pippus de Brunelleschis de magnifica et potente civitate Florentie / honore egregius ac famosus in pluribus virtutibus a deo dotatus et maxime in architectura / in designoque inventor de edifitiis prespicacissimus / ex sua be-

[15] From Guasti, *Belle arti*. Translation: p. 118.
[16] Literally this may mean: "[A man of] false judgment loses such hope, for experience is terrible to him." However, the sense is even less clear than in Acquettini's production.
[17] A possible literal meaning is the following: "For the wise man nothing is invisible, except that which does not exist, since he has none of the idle dreams of one not really [as asserted by Acquettini] of the school." The text, while no doubt accurately copied by Guasti, appears very confused in his publication. It would end a sentence with the incredibly trivial phrase, *quel che non è perche ha mancanza*. It would then continue with a meaningless phrase, beginning: *Et quelle fantasie*. The confusion is compounded by a reference to *un non si sicola*. *Sanpaolesi* recognized that the last word or words must mean *di scuola*.
[18] Probable literal meaning: "The essence of that which nature hides is given [interpreted] by art. It is not seen by every [one who engages in] false thought."
[19] From Lat. 197, fols. 107ᵛ–109ʳ, Munich.

nignitate in Sena dixit mihi hoc verbo / Noli cum multis participare inventiones tuas / secus paucis inteligentibus / et amatoribus scientiarum quia nimis ostendere et dicere suas inventiones et facta sua / est unum [tantum quod] derogare sua ingenia. Multi sunt [qui] libentius audiunt causa inventores vituperandi [et] opponendi suis factis et dictis / causa ne in aliquo honorato loco audiantur / Et postea per aliquot menses aut annum dicunt eadem verba oretenus / aut in scripturis aut in designis / et dicunt auda[c]ter se fuisse inventores earum rerum / de quibus ante fuit male locutus / et gloriam alterius sibi tribuit / Et alter est materialis et grossi ingenei / et audit res novas et ingenia numque auditas miratur valde et deridit de inventore / et sui ingenii / et dicens inventori fac mihi hoc servitium ne dicas ista talia plus / tu reputaveris una bestia. Unde pro malis dicentibus ex invidia motis aut ingnorantia motis / non debentur dona dei tributa relinquere sed ea sequi ac exercere / qui[bus] virtuosi ac ingeniosi a sapientioribus sapientes reputantur enim

(Fo. 108ᵛ) Omnibus et singulis non potest demostrari secretrum aque maris / aut fluminis et eorum edifitia / sed adunata sinodo sapientium filosophorum / ac magistrorum / in arte machanica expertorum delibarant omne id necessarium / in opare conponendo ac edificando / unus quisque vult scire de materia proponenda tam doctus quam indoctus / doctius propostam materiam oparis sive edifitii / inteligit et semper aliquid inteligit / aut mediocriter sive totaliter / sed indocti ac imperiti / nichil inteligunt / ut quando eis assignatur ratio quam ipsi non inteligunt statim ab ingnorantia eorum moventur ad iram / stant costantes in eorum ingnorantia / quia volunt se ostendere scientes et non sunt / imo ducunt alios ingnorantes ad fortificandum sua mala dicta / ac dicentes inteligentibus contumeliam / Unde est maximum periculum coram capocchis et ingnorantibus de aquistuctis et stringendis aquis ac asscendentibus et descendentibus tam sub terra quam super terram / et de edifitiis in aquis sive super aquam tam marinam quam dulcem / Et isti tales scientes sunt valde diligendi et ingnorantes fugiendi multo magis. ingnorantes capitosique ad praelium sunt mictendi / scientes autem ad consilium quia sunt honor et gloria Reiplubice. Amen.

Nota che due cose sonno quelle che sonno malagevogli ne fiumi reagli quando vogli edificare stechaie da mulino / overo edificare ponti / sonno queste due prima exvollare el fiume / che non levi via e lavorio / che tu fai / la seconda sie fare el guazatoio / per modo che non roda le coste del ponte / overo el muro et fondamento della stechaia murata / La terza se non poi exvollare el f[i]ume di suo lecto / lassare una parte della stechaia vota / et non murata acio chel fiume inde possa passare / acio che non inpedischa el tuo edifitio / et ordinare con caduta / o cateracta serrare el detto f[i]ume / la quarta cosa e di vedere sel fiume ane el suo lecto / di tufo pietra grande / et se que[s]te cose ane el detto fiume / Non si vuole ficcari in e pagli / inperoche pagli franghono el leto del fiume / et

facto ledifitio venendo diluvio daqua truova el terreno rocto dai pagli / el diluvio el porta secho / et rimane ledifitio in su pagli / et corre laque sotto e fondamenti / Et del fondamento et soddo murare sopra el detto fondamento del leto del fiume.

Gli stechoni funo trovati quando el fiume ane el fondo debile / overo luoghi renosi / o aqua gemogli / o solfinaie / et in questi luoghi si ficchano e pagli / sopra de quagli si fondano le chose de ponti overo e muri delle stecchaie / Laltra cosa sie quando el muro e facto di frescho sapello covertare di legname acio che laqua del fiume non levi via el tuo lavorio / Et prima in que lugho farvi la cappana per conservare la cantina et massaritie de maestri Et per habitatione di loro persone etc.

Et nota se fai edifi[t]io in fiume come di sopra e detto prima guarda se apresso ve boschi o luoghi da ffare e legname et se ve presso pietre da murare o da fare calcina / et anche ghia[ia] / o rena da murare / et se queste cose non sonno presso al detto luogho / chi in tal cosa si pactara / o tollara a ffare / vorra di richo povaro diventara / Et savio tengho chi non si pactara perche la spesa de legname / pietre / calcina fornaci atrainare somegg[i]are / carregare et manifacture et logro della vita / che di tute queste cose non si puo fare stima etc. Se tollilo a ffare ordinari da colui che fare vole lavorio etc.

145

BRUNELLESCHI'S INFLUENCE

Of the men listed in the opposite chart, Mariano Taccola appears as a most influential follower of Filippo the engineer. Taccola's followers produced a number of straightforward copies, usually taken from the *De ingeneis* of 1433. These copies are mixed with material from a Machinery Complex, of unknown origin,[20] but they are also rich in Brunelleschian recollections.

In the later Quattrocento, Francesco di Giorgio produced an important version of Taccola's work, in his *De architectura* and a pocket-sized book of fairly similar content.[21] In his later *Trattati* he repeated most of this material, adding a few mechanical devices not previously noted, a variety of architectural plans, and specific Vitruvian ideas.[22] These ideas appeared here for the first time, so far as machinery is concerned; Vitruvian influence had previously entered only the theoretical architectural writings or translations of Ghiberti, Alberti, and Fra Giocondo. The various elements then were further elaborated by anonymous assistants and followers of Francesco,[23] by the Biringuccios,[24] P. Cataneo,[25] and authors of early printed books. Some Brunelleschian inventions may still be found in these later productions.[26]

Other followers of Taccola contributed in similar form, although with less originality of a technical kind and less perseverance in the rediscovery of ancient knowledge. The works of such followers exist in considerable numbers.[27] In general they

[20] We postulate this Machinery Complex on the basis of identical copies in *De architectura* by Francesco di Giorgio and manuscripts independent of him (*infra*, note 27).

[21] Harley 3281 in British Museum and Urb. Lat. 1757 in Vaticana; see A. E. Popham and P. Pouncey, *Italian Drawings in . . . the British Museum*, London, 1950, pp. 33, 36, and the expected publication of the Urbinatus by L. Michelini Tocci, Rome. Copies of the Harleianus: Turin, N.L. 383 and Modena, Ital. 421.

[22] Originals of many are in the Uffizi at Florence. Straight copies from *De ingeneis* may be found in major parts of *Trattati*, Pls. 11–15, 40, 41, 47–49, 52, 78, 82, 89, 100–102, 110–120, 124–126.

[23] Florence, Accademie, E. 2. I. 28 and Biblioteca Nazionale, Magl. I. II. 141; Siena, S. IV. 4. About these see A. Parronchi, "Di un manoscritto attribuito a Francesco di Giorgio Martini," *La colombaria*, XXXI, 1966, pp. 165–213.

[24] Vannoccio Biringuccio, *Pirotechnia*, Venice, 1540; Oreste Biringuccio, unpublished manuscript, Siena, S. IV. 1, of the late Cinquecento.

[25] "*Il libro di P. Cataneo*," manuscript in Uffizi at Florence, partly published in Cataneo's *L'architettura*, Venice, 1554 and a later edition. Folio 1 of the book published in 1554 has an interesting variant of Palat. 766, 41v, Florence, Bibl. Naz. referring specially to perspective: *Sarà di bisogno . . . [l'architetto] essere scientifico e di naturale ingegno dotato . . . essere bono . . . prospettivo . . . Se l'architetto non serà prospettivo non potra mai così bene honorarsi.* A Brunelleschian addition to Taccola's Brunelleschian recollection?

[26] Lorini, Branca, Zonca, and others.

[27] Including Florence, Bibl. Naz. Palat. 767 and its copy, Venice, Ital. Z. 86, published in part by G. Canestrini, *Arte militare meccanica medievale*, Milan, 1946; and the unpublished manuscripts at London, Brit. Mus., Add. 34113; Uffizi, Florence, Collection Antonio da San Gallo il Giovane; Bibl. Naz., Florence, Magl. II. III. 314 (formerly XVIII. 4) and XVIII. V. 2; Milano, Ambrosiana, N. 191.

Filippo   Lorenzo   1400
Brunelleschi   Ghiberti

Mariano
Taccola   Filarete

Fra   Alberti
Giocondo

Francesco   Giuliano   Buonaccorso
di Giorgio   San Gallo   Ghiberti   1500

Leonardo   Antonio
da Vinci   San Gallo   Serlio

Michel
Angelo   P. Cataneo   Palladio

G. della   Domenico O. Biringuccio
Porta   Fontana
                                                       1600
                      B. Lorini

                      V. Zonca

                      G. Branca

Carlo   "Theaters
Fontana   of Machines"   1700
Nelli

            "Tre
            Matematici"

            Poleni

                      Modern Textbooks of
                      Civil and Mechanical
                      Engineering and   1800
                      Architecture

illustrate machines that also appear in the works of Francesco di Giorgio,[28] and sometimes they show architectures similar to his. Occasionally they add a small architectural scheme of Brunelleschi's time.[29] At some points, as we have shown, they mention the construction of the Cupola; elsewhere they refer to the names of recent engineers, but they seem unaware of Filippo.[30] Perhaps we should say they seem unaware of a reason for mentioning Filippo.

Yet the influence of Brunelleschi's technical work remains evident in these manuscripts. This influence is just as evident as the improvement in perspective rendering, and the effects of Filippo's architectural innovations. The chart summarizes names and approximate positions of the main followers recognizable at this time. Needless to say, all names and lines of transmission of ideas are in need of further review.

[28] For example, drawings of clock developments, perhaps of Brunelleschian or contemporary origin. Prager, "Brunelleschi's Clock?" pp. 206, 211 f.

[29] For example Siena, S. IV. 6, fo. 34^r.

[30] Venice, Lat. 2941, fo. 77^r mentions Francesco di Giorgio, as other manuscripts do. A nineteenth-century note in front of Lat. 2941 also mentions that the text attributes some machines to Filippo, *a . . . Philippo machinatore sumptas esse nonnullas adscriptum est.* We found no such text.

A CATALOG OF SELECTED DOVER BOOKS IN ALL FIELDS OF INTEREST

100 BEST-LOVED POEMS, Edited by Philip Smith. "The Passionate Shepherd to His Love," "Shall I compare thee to a summer's day?" "Death, be not proud," "The Raven," "The Road Not Taken," plus works by Blake, Wordsworth, Byron, Shelley, Keats, many others. 96pp. 5³⁄₁₆ x 8¼. 0-486-28553-7

100 SMALL HOUSES OF THE THIRTIES, Brown-Blodgett Company. Exterior photographs and floor plans for 100 charming structures. Illustrations of models accompanied by descriptions of interiors, color schemes, closet space, and other amenities. 200 illustrations. 112pp. 8⅜ x 11. 0-486-44131-8

1000 TURN-OF-THE-CENTURY HOUSES: With Illustrations and Floor Plans, Herbert C. Chivers. Reproduced from a rare edition, this showcase of homes ranges from cottages and bungalows to sprawling mansions. Each house is meticulously illustrated and accompanied by complete floor plans. 256pp. 9⅜ x 12¼. 0-486-45596-3

101 GREAT AMERICAN POEMS, Edited by The American Poetry & Literacy Project. Rich treasury of verse from the 19th and 20th centuries includes works by Edgar Allan Poe, Robert Frost, Walt Whitman, Langston Hughes, Emily Dickinson, T. S. Eliot, other notables. 96pp. 5³⁄₁₆ x 8¼. 0-486-40158-8

101 GREAT SAMURAI PRINTS, Utagawa Kuniyoshi. Kuniyoshi was a master of the warrior woodblock print — and these 18th-century illustrations represent the pinnacle of his craft. Full-color portraits of renowned Japanese samurais pulse with movement, passion, and remarkably fine detail. 112pp. 8⅜ x 11. 0-486-46523-3

ABC OF BALLET, Janet Grosser. Clearly worded, abundantly illustrated little guide defines basic ballet-related terms: arabesque, battement, pas de chat, relevé, sissonne, many others. Pronunciation guide included. Excellent primer. 48pp. 4³⁄₁₆ x 5¾. 0-486-40871-X

ACCESSORIES OF DRESS: An Illustrated Encyclopedia, Katherine Lester and Bess Viola Oerke. Illustrations of hats, veils, wigs, cravats, shawls, shoes, gloves, and other accessories enhance an engaging commentary that reveals the humor and charm of the many-sided story of accessorized apparel. 644 figures and 59 plates. 608pp. 6 ⅛ x 9¼. 0-486-43378-1

ADVENTURES OF HUCKLEBERRY FINN, Mark Twain. Join Huck and Jim as their boyhood adventures along the Mississippi River lead them into a world of excitement, danger, and self-discovery. Humorous narrative, lyrical descriptions of the Mississippi valley, and memorable characters. 224pp. 5³⁄₁₆ x 8¼. 0-486-28061-6

ALICE STARMORE'S BOOK OF FAIR ISLE KNITTING, Alice Starmore. A noted designer from the region of Scotland's Fair Isle explores the history and techniques of this distinctive, stranded-color knitting style and provides copious illustrated instructions for 14 original knitwear designs. 208pp. 8⅜ x 10⅞. 0-486-47218-3

Browse over 9,000 books at www.doverpublications.com

ALICE'S ADVENTURES IN WONDERLAND, Lewis Carroll. Beloved classic about a little girl lost in a topsy-turvy land and her encounters with the White Rabbit, March Hare, Mad Hatter, Cheshire Cat, and other delightfully improbable characters. 42 illustrations by Sir John Tenniel. 96pp. 5³⁄₁₆ x 8¼. 0-486-27543-4

AMERICA'S LIGHTHOUSES: An Illustrated History, Francis Ross Holland. Profusely illustrated fact-filled survey of American lighthouses since 1716. Over 200 stations — East, Gulf, and West coasts, Great Lakes, Hawaii, Alaska, Puerto Rico, the Virgin Islands, and the Mississippi and St. Lawrence Rivers. 240pp. 8 x 10¾.

0-486-25576-X

AN ENCYCLOPEDIA OF THE VIOLIN, Alberto Bachmann. Translated by Frederick H. Martens. Introduction by Eugene Ysaye. First published in 1925, this renowned reference remains unsurpassed as a source of essential information, from construction and evolution to repertoire and technique. Includes a glossary and 73 illustrations. 496pp. 6⅛ x 9¼. 0-486-46618-3

ANIMALS: 1,419 Copyright-Free Illustrations of Mammals, Birds, Fish, Insects, etc., Selected by Jim Harter. Selected for its visual impact and ease of use, this outstanding collection of wood engravings presents over 1,000 species of animals in extremely lifelike poses. Includes mammals, birds, reptiles, amphibians, fish, insects, and other invertebrates. 284pp. 9 x 12. 0-486-23766-4

THE ANNALS, Tacitus. Translated by Alfred John Church and William Jackson Brodribb. This vital chronicle of Imperial Rome, written by the era's great historian, spans A.D. 14-68 and paints incisive psychological portraits of major figures, from Tiberius to Nero. 416pp. 5³⁄₁₆ x 8¼. 0-486-45236-0

ANTIGONE, Sophocles. Filled with passionate speeches and sensitive probing of moral and philosophical issues, this powerful and often-performed Greek drama reveals the grim fate that befalls the children of Oedipus. Footnotes. 64pp. 5³⁄₁₆ x 8 ¼. 0-486-27804-2

ART DECO DECORATIVE PATTERNS IN FULL COLOR, Christian Stoll. Reprinted from a rare 1910 portfolio, 160 sensuous and exotic images depict a breathtaking array of florals, geometrics, and abstracts — all elegant in their stark simplicity. 64pp. 8⅜ x 11. 0-486-44862-2

THE ARTHUR RACKHAM TREASURY: 86 Full-Color Illustrations, Arthur Rackham. Selected and Edited by Jeff A. Menges. A stunning treasury of 86 full-page plates span the famed English artist's career, from *Rip Van Winkle* (1905) to masterworks such as *Undine, A Midsummer Night's Dream,* and *Wind in the Willows* (1939). 96pp. 8⅜ x 11.

0-486-44685-9

THE AUTHENTIC GILBERT & SULLIVAN SONGBOOK, W. S. Gilbert and A. S. Sullivan. The most comprehensive collection available, this songbook includes selections from every one of Gilbert and Sullivan's light operas. Ninety-two numbers are presented uncut and unedited, and in their original keys. 410pp. 9 x 12.

0-486-23482-7

THE AWAKENING, Kate Chopin. First published in 1899, this controversial novel of a New Orleans wife's search for love outside a stifling marriage shocked readers. Today, it remains a first-rate narrative with superb characterization. New introductory Note. 128pp. 5³⁄₁₆ x 8¼. 0-486-27786-0

BASIC DRAWING, Louis Priscilla. Beginning with perspective, this commonsense manual progresses to the figure in movement, light and shade, anatomy, drapery, composition, trees and landscape, and outdoor sketching. Black-and-white illustrations throughout. 128pp. 8⅜ x 11. 0-486-45815-6

THE BATTLES THAT CHANGED HISTORY, Fletcher Pratt. Historian profiles 16 crucial conflicts, ancient to modern, that changed the course of Western civilization. Gripping accounts of battles led by Alexander the Great, Joan of Arc, Ulysses S. Grant, other commanders. 27 maps. 352pp. 5⅜ x 8½. 0-486-41129-X

BEETHOVEN'S LETTERS, Ludwig van Beethoven. Edited by Dr. A. C. Kalischer. Features 457 letters to fellow musicians, friends, greats, patrons, and literary men. Reveals musical thoughts, quirks of personality, insights, and daily events. Includes 15 plates. 410pp. 5⅜ x 8½. 0-486-22769-3

BERNICE BOBS HER HAIR AND OTHER STORIES, F. Scott Fitzgerald. This brilliant anthology includes 6 of Fitzgerald's most popular stories: "The Diamond as Big as the Ritz," the title tale, "The Offshore Pirate," "The Ice Palace," "The Jelly Bean," and "May Day." 176pp. 5⅜ x 8½. 0-486-47049-0

BESLER'S BOOK OF FLOWERS AND PLANTS: 73 Full-Color Plates from Hortus Eystettensis, 1613, Basilius Besler. Here is a selection of magnificent plates from the *Hortus Eystettensis*, which vividly illustrated and identified the plants, flowers, and trees that thrived in the legendary German garden at Eichstätt. 80pp. 8⅜ x 11.
0-486-46005-3

THE BOOK OF KELLS, Edited by Blanche Cirker. Painstakingly reproduced from a rare facsimile edition, this volume contains full-page decorations, portraits, illustrations, plus a sampling of textual leaves with exquisite calligraphy and ornamentation. 32 full-color illustrations. 32pp. 9⅜ x 12¼. 0-486-24345-1

THE BOOK OF THE CROSSBOW: With an Additional Section on Catapults and Other Siege Engines, Ralph Payne-Gallwey. Fascinating study traces history and use of crossbow as military and sporting weapon, from Middle Ages to modern times. Also covers related weapons: balistas, catapults, Turkish bows, more. Over 240 illustrations. 400pp. 7¼ x 10⅛. 0-486-28720-3

THE BUNGALOW BOOK: Floor Plans and Photos of 112 Houses, 1910, Henry L. Wilson. Here are 112 of the most popular and economic blueprints of the early 20th century — plus an illustration or photograph of each completed house. A wonderful time capsule that still offers a wealth of valuable insights. 160pp. 8⅜ x 11.
0-486-45104-6

THE CALL OF THE WILD, Jack London. A classic novel of adventure, drawn from London's own experiences as a Klondike adventurer, relating the story of a heroic dog caught in the brutal life of the Alaska Gold Rush. Note. 64pp. 5³⁄₁₆ x 8¼.
0-486-26472-6

CANDIDE, Voltaire. Edited by Francois-Marie Arouet. One of the world's great satires since its first publication in 1759. Witty, caustic skewering of romance, science, philosophy, religion, government — nearly all human ideals and institutions. 112pp. 5³⁄₁₆ x 8¼. 0-486-26689-3

CELEBRATED IN THEIR TIME: Photographic Portraits from the George Grantham Bain Collection, Edited by Amy Pastan. With an Introduction by Michael Carlebach. Remarkable portrait gallery features 112 rare images of Albert Einstein, Charlie Chaplin, the Wright Brothers, Henry Ford, and other luminaries from the worlds of politics, art, entertainment, and industry. 128pp. 8⅜ x 11. 0-486-46754-6

CHARIOTS FOR APOLLO: The NASA History of Manned Lunar Spacecraft to 1969, Courtney G. Brooks, James M. Grimwood, and Loyd S. Swenson, Jr. This illustrated history by a trio of experts is the definitive reference on the Apollo spacecraft and lunar modules. It traces the vehicles' design, development, and operation in space. More than 100 photographs and illustrations. 576pp. 6¾ x 9¼. 0-486-46756-2

A CHRISTMAS CAROL, Charles Dickens. This engrossing tale relates Ebenezer Scrooge's ghostly journeys through Christmases past, present, and future and his ultimate transformation from a harsh and grasping old miser to a charitable and compassionate human being. 80pp. 5³⁄₁₆ x 8¼. 0-486-26865-9

COMMON SENSE, Thomas Paine. First published in January of 1776, this highly influential landmark document clearly and persuasively argued for American separation from Great Britain and paved the way for the Declaration of Independence. 64pp. 5³⁄₁₆ x 8¼. 0-486-29602-4

THE COMPLETE SHORT STORIES OF OSCAR WILDE, Oscar Wilde. Complete texts of "The Happy Prince and Other Tales," "A House of Pomegranates," "Lord Arthur Savile's Crime and Other Stories," "Poems in Prose," and "The Portrait of Mr. W. H." 208pp. 5³⁄₁₆ x 8¼. 0-486-45216-6

COMPLETE SONNETS, William Shakespeare. Over 150 exquisite poems deal with love, friendship, the tyranny of time, beauty's evanescence, death, and other themes in language of remarkable power, precision, and beauty. Glossary of archaic terms. 80pp. 5³⁄₁₆ x 8¼. 0-486-26686-9

THE COUNT OF MONTE CRISTO: Abridged Edition, Alexandre Dumas. Falsely accused of treason, Edmond Dantès is imprisoned in the bleak Chateau d'If. After a hair-raising escape, he launches an elaborate plot to extract a bitter revenge against those who betrayed him. 448pp. 5³⁄₁₆ x 8¼. 0-486-45643-9

CRAFTSMAN BUNGALOWS: Designs from the Pacific Northwest, Yoho & Merritt. This reprint of a rare catalog, showcasing the charming simplicity and cozy style of Craftsman bungalows, is filled with photos of completed homes, plus floor plans and estimated costs. An indispensable resource for architects, historians, and illustrators. 112pp. 10 x 7. 0-486-46875-5

CRAFTSMAN BUNGALOWS: 59 Homes from "The Craftsman," Edited by Gustav Stickley. Best and most attractive designs from Arts and Crafts Movement publication — 1903–1916 — includes sketches, photographs of homes, floor plans, descriptive text. 128pp. 8¼ x 11. 0-486-25829-7

CRIME AND PUNISHMENT, Fyodor Dostoyevsky. Translated by Constance Garnett. Supreme masterpiece tells the story of Raskolnikov, a student tormented by his own thoughts after he murders an old woman. Overwhelmed by guilt and terror, he confesses and goes to prison. 480pp. 5³⁄₁₆ x 8¼. 0-486-41587-2

THE DECLARATION OF INDEPENDENCE AND OTHER GREAT DOCUMENTS OF AMERICAN HISTORY: 1775-1865, Edited by John Grafton. Thirteen compelling and influential documents: Henry's "Give Me Liberty or Give Me Death," Declaration of Independence, The Constitution, Washington's First Inaugural Address, The Monroe Doctrine, The Emancipation Proclamation, Gettysburg Address, more. 64pp. 5³⁄₁₆ x 8¼. 0-486-41124-9

THE DESERT AND THE SOWN: Travels in Palestine and Syria, Gertrude Bell. "The female Lawrence of Arabia," Gertrude Bell wrote captivating, perceptive accounts of her travels in the Middle East. This intriguing narrative, accompanied by 160 photos, traces her 1905 sojourn in Lebanon, Syria, and Palestine. 368pp. 5⅜ x 8½. 0-486-46876-3

A DOLL'S HOUSE, Henrik Ibsen. Ibsen's best-known play displays his genius for realistic prose drama. An expression of women's rights, the play climaxes when the central character, Nora, rejects a smothering marriage and life in "a doll's house." 80pp. 5³⁄₁₆ x 8¼. 0-486-27062-9

Browse over 9,000 books at www.doverpublications.com